Nanoscale Hydrodynamics of Simple Systems

Written for graduate students and researchers, *Nanoscale Hydrodynamics of Simple Systems* covers fundamental aspects of nanoscale hydrodynamics and gives examples of its applications. Covering classical, generalised, and extended hydrodynamic theories, the title also discusses their limitations. It introduces the reader to nanoscale fluid phenomena and explores how fluid dynamics on this extreme length scale can be understood using hydrodynamic theory and detailed atomistic simulations. It also comes with additional resources, including a series of explanatory videos on the installation of a simulation software package, as well as discussion, analysis, and visualisations of simulations. This title primarily focuses on training the reader to identify when classical theory breaks down, how to extend and generalise the theory, as well as assimilate how simulations and theory together can be used to gain fundamental knowledge about the fluid dynamics of small-scale systems.

Jesper Schmidt Hansen is Professor in Integrated Science at the Department of Science and Environment, Roskilde University, Denmark. His research focus is nanoscale fluid systems explored through molecular dynamics and hydrodynamics. He has authored more than 50 original research articles and has taught many courses ranging from C-programming and mathematical analysis to quantum mechanics.

Nanoscale Hydrodynamics
of Simple Systems

Jesper Schmidt Hansen

Roskilde University

CAMBRIDGE
UNIVERSITY PRESS

CAMBRIDGE
UNIVERSITY PRESS

University Printing House, Cambridge CB2 8BS, United Kingdom

One Liberty Plaza, 20th Floor, New York, NY 10006, USA

477 Williamstown Road, Port Melbourne, VIC 3207, Australia

314–321, 3rd Floor, Plot 3, Splendor Forum, Jasola District Centre, New Delhi – 110025, India

103 Penang Road, #05–06/07, Visioncrest Commercial, Singapore 238467

Cambridge University Press is part of the University of Cambridge.

It furthers the University's mission by disseminating knowledge in the pursuit of
education, learning, and research at the highest international levels of excellence.

www.cambridge.org
Information on this title: www.cambridge.org/9781009158732
DOI: 10.1017/9781009158749

First published 2022

A catalogue record for this publication is available from the British Library.

Library of Congress Cataloging-in-Publication Data
Names: Hansen, Jesper Schmidt, 1973- author.
Title: Nanoscale hydrodynamics of simple systems / Jesper Schmidt Hansen.
Description: New York : Cambridge University Press, 2022. | Includes
bibliographical references and index.
Identifiers: LCCN 2022002801 (print) | LCCN 2022002802 (ebook) | ISBN
9781009158732 (hardback) | ISBN 9781009158749 (ebook)
Subjects: LCSH: Hydrodynamics. | Nanofluids. | BISAC: SCIENCE / Physics /
General
Classification: LCC QC151 .H35 2022 (print) | LCC QC151 (ebook) | DDC
532/.05–dc23/eng20220524
LC record available at https://lccn.loc.gov/2022002801
LC ebook record available at https://lccn.loc.gov/2022002802

ISBN 978-1-009-15873-2 Hardback

Additional resources for this publication at www.cambridge.org/nanoscale.

Contents

Preface

Nanotechnology is referred to as a new emerging technology. The reason why nanotechnology is labelled "new" is linked to fascinating advances in novel experimental techniques, which today allow for controllable fabrication of nanoscale ducts, fluid channels, and chambers. These nanoscale geometries offer new methods and devices for chemical separation techniques, fast reactions, and energy conversions.

To fully understand how nanoscale devices operate, and thereby how to control and optimise the technology, theoretical models act as a key component. Importantly, we should not expect that nanotechnology is based on fundamental new physics, but rather that we can apply already established theoretical frameworks. This does not mean that nanoscale modelling is trivial, and frequently one encounters phenomena that are not observed on the macroscopic length scale. Nevertheless, the observation can often be modelled from well-known theories. Richard Feynman's famous quote 'There is plenty of room at the bottom' from 1959 is often referred to as the start of nanotechnology, and at least from a modelling point of view, nanotechnology is perhaps not that new.

This book specifically explores nanoscale hydrodynamics of simple systems. 'Simple systems' here refers to simple straight channel geometries and fluids composed of molecules like methane, butane, and water. As evident from the focus statement – and book title – the theoretical framework is based on hydrodynamics. The basic exploration workflow is to compare the hydrodynamic model predictions and data from atomistic computer simulations, which enables a very detailed discussion. Atomistic computer simulations are widely used today to investigate fluid and liquid dynamics and structure and are made possible by both efficient algorithms and also improved hardware performance. This scientific method is relatively new. Of course, the final test for any theory must be based on physical experiments.

The book aims to give the reader an in-depth understanding of hydrodynamic modelling of fluids and liquids on the nanoscale. To achieve this goal the reader will learn

1. about different nanoscale hydrodynamic phenomena,
2. how hydrodynamic quantities are defined from the molecular quantities,
3. how the hydrodynamic equations can be formulated from these definitions,
4. how to analyse the hydrodynamic equations and compare the analysis results with simulation data,
5. when and how classical hydrodynamic theory breaks down,

6. how hydrodynamics can be extended in order to model and understand phenomena, where classical theory breaks down, and

7. about the synergy between simulations and theory.

The learning outcome is supported by the 'Further Explorations' sections at the end of Chapters 2–6. The explorations are a mixture of short, well-defined exercises and more open research problems. The open problems often require computer simulations, and to this end the reader (well, in fact, everyone) has access to computer resources in the form of a molecular dynamics kernel library and a GNU Octave/Matlab interface. Some of the simulations can be long, depending on the hardware and the problem, and here a bit of patience can be required. The software does not need supercomputer facilities or an IT specialist to install, and can run from a standard desktop or a fast laptop; a Linux operating system or a Linux emulator is strongly recommended. From the book web page a series of videos is available showing how to install and use the software.

The book is written for anyone who wishes to pursue research on the topic, or who is simply curious about hydrodynamics on the nanoscale. The text difficulty is chosen such that some prior knowledge of linear algebra, vector calculus, electrodynamics, and hydrodynamics is required. Concepts like tensors and the outer product will be briefly introduced, as these may not be known to the reader. These introductions are meant to enable the reader to move on to the next part of the text, and are not, by any standard, thorough treatments. Moreover, it is my experience that formulating the relevant differential equation for a given problem is usually tractable, but the techniques for solving the equation can pose problems. Fortunately, nanoscale hydrodynamics is mostly almost linear, and the differential equations we encounter are relatively simple. In any case, I have chosen to explain the steps involved in solving these equations carefully. Finally, the Appendix contains some clarification regarding the symbolism used in the book and also literature suggestions for further reading. In case the reader comes across an unknown concept, an abundance of good resources can be found on the internet.

Despite the help I have received in writing this book, there will be errors; only the author is to blame for these, and I hope for the readers' forgiveness. Also, some may find some topics and literature missing. The book is not in any way meant as a prioritisation of all the great work done by the community.

I would like to thank Bjarke Spangsberg Bak for going through the text minutely, weeding out numerous typos, sign errors, and even finding calculus mistakes. I thank Thomas Voigtmann for his many insightful comments and suggestions, Lorenzo Costigliola, and Solvej Knudsen for reading parts of the manuscript at the early stage of writing. A special thank-you goes out to Jeppe Dyre for his encouragement over the years, and also for the financial support for this book (VILLUM Foundation's *Matter* grant – No. 16515). Finally, I must thank my family for their sustained and loving support.

Symbol List

α	Thermal gradient or eigenvalue
$\mathcal{A}$	Operator representation of Maxwell's constitutive model
a_i	General molecular quantity
A	General hydrodynamic variable or surface area
$A_{\mathrm{av}}, \delta A$	Average and fluctuating parts of A
$\overset{s}{\mathbf{A}}, \overset{os}{\mathbf{A}}$	Symmetric and traceless symmetric part of tensor $\mathbf{A}$
$\overset{a}{\mathbf{A}}, \overset{ad}{\mathbf{A}}$	Antisymmetric part and vector dual of tensor $\mathbf{A}$
$\langle A \rangle$	Ensemble average of a set of independent measurements $A_1, A_2 \ldots$
$\langle A \rangle_t$	Time average of A
$\overline{A}$	Spatial average of A
$\widetilde{A}$	Fourier coefficient
$\widehat{A}$	Fourier–Laplace coefficient
A_{rz}	Reaction area normal to wavefront propagation
β_V	Thermal pressure coefficient
β_T	Isothermal compressibility
$\mathbf{c}_i$	Thermal (peculiar) velocity of molecule i
C	General correlation function
$C^{\perp}, C^{\parallel}$	Transverse and longitudinal correlation functions
C_{xy}	xy-correlation function
c_V, c_P	Specific heat capacities at constant volume and constant pressure
c_T	Transverse shear wave speed
c_{scr}	Charge screening function
$c_{\mathrm{min}}, c_{\mathrm{rel}}$	Minimum and relative chemical wavefront speeds
Δ	Slab width of fluid element adjacent to wall or difference
$\Delta\omega_{\mathrm{Ra}}$	Rayleigh half-peak width
δ_S	Stokes boundary layer for oscillatory flows
$\mathbf{D}$	Electric displacement field
$D_s, \mathbf{D}_s$	Self-diffusion coefficient and self-diffusion tensor
D_A	Self-diffusion coefficient of molecule A
D_{AB}	Mutual diffusion coefficient of molecule A in A-B mixture

D_T	Thermal diffusivity or thermal diffusion coefficient
De	Deborah number
ε	Lennard–Jones energy scale or dielectric permittivity
$\varepsilon_0, \varepsilon_r$	Dielectric permittivity in vacuum and relative dielectric permittivity
ε	Small parameter/perturbation parameter
ε_i	Thermal kinetic energy of molecule i
η_0, η_v, η_r	Shear, bulk, and rotational viscosities
η_t	Transverse viscosity $\eta_t = \eta_0 + \eta_r$
E	Flow enhancement coefficient
$\mathbf{E}, \mathbf{E}^{\text{ext}}$	Local and external electric fields
$\mathcal{F}$	The Fourier transform operator
f	Spatial part of transport kernel or generic function
F, F_s	The coherent and incoherent intermediate scattering functions
$\mathbf{F}_i$	Force acting on molecule i
$\mathbf{F}_{ij}$	Force acting on molecule i due to j
$\mathbf{F}^c, \mathbf{F}^{\text{ext}}, \mathbf{F}^{\text{tot}}$	Conservative, external, and total forces
$\mathbf{F}^D, \mathbf{F}^R$	Dissipative and random forces in DPD simulations
γ	Ratio of heat capacities
$\dot{\boldsymbol{\gamma}}, \dot{\gamma}$	Strain rate tensor and tensor component
Γ	Sound attenuation coefficient
$\mathcal{G}$	The general microscopic operator
g	Radial distribution function
$\mathbf{g}$	Applied force per unit mass (applied acceleration).
G, G_s	The van Hove function and self-part of the van Hove function
G_∞	Modulus of rigidity
$\mathcal{H}$	The microscopic hydrodynamic operator
$h, \bar{h}$	Slit-pore height and half-height
$\mathbf{j}$	Momentum density
$\mathbf{J}$	General flux tensor
$\mathbf{J}^i, \mathbf{J}^A$	Single-particle flux tensor
$\mathbf{J}^\varepsilon$	Thermal kinetic energy flux
$\mathbf{J}^L$	Orbital momentum flux
χ_e	Electric susceptibility
κ	$\kappa = \lambda/(c_V \rho_{\text{av}})$
k	Chemical reaction rate constant
$\mathbf{k}$	Wavevector, $\mathbf{k} = (k_x, k_y, k_z)$
K_1, K_2, K_3	Integration constants
λ_D	Debye length
$\mathcal{L}$	The Fourier–Laplace transform operator
l	Slit-pore length
l_c	Characteristic length scale for momentum coupling effect
$\mathbf{L}_i$	Orbital momentum of molecule i

L_K	Kapitza length
$L_s, L_s^{(1)}, L_s^{(2)}$	Slip length and slip lengths at wall 1 and wall 2
v_l	Longitudinal viscosity $v_l = (\eta_v + 4\eta_0/3)/\rho_{av}$
n, n_A	Number density and number density of molecule A
n_{rz}	Number of molecules in reaction zone
$\mathbf{n}$	Normal vector
$\mathbf{N}, \mathbf{N}_i$	Torque and torque on molecule i
∇	The del-operator
$\boldsymbol{\mu}_i$	Molecular/microscopic dipole
m_i	Mass of molecule i
M	Constant of proportionality (transport coefficient)
$\mathbf{M}, \mathbf{M}_i$	Torque, and torque on molecule i with respect to c.o.m.
Ma	Mach number
Re	Reynolds number
$\tau_D, \tau_F,$	Debye relaxation time, Frenkel escape time, Maxwell relaxation
τ_M, τ_s	time, and stress relaxation time
τ_{obs}	Observation time
τ_{path}	Characteristic time between particle momentum exchange
ρ	Mass density
$\rho\varepsilon$	Thermal kinetic energy density
$\rho\mathbf{L}, \rho\mathbf{S}$	Orbital and spin angular momentum densities
$\mathbf{u}$	Fluid streaming velocity, $\mathbf{u} = (u_x, u_y, u_z)$
u_w	Velocity at wall
ω	Frequency or eigenvalue
ω_{peak}	Peak frequency
$\boldsymbol{\Omega}, \boldsymbol{\Omega}_i$	Spin angular velocity and spin angular velocity of molecule i
Π	Viscous pressure
ϕ	Associated field variable per unit mass, or kernel, or auxiliary function
Φ_T	Thiele modulus
φ_c, φ_q	Boltzmann and electric potential functions
$\mathbf{p}_i$	Momentum of molecule i
p_{eq}	Equilibrium pressure
$\mathbf{P}$	Polarization (rank-1 tensor) or pressure tensor (rank-2 tensor)
Pr	Prandtl number
q, q_{scr}	Charge and screening charge
Q	Volumetric flow rate
$\mathbf{Q}$	Spin angular momentum flux tensor
ρ_q, ρ_b, ρ_f	Charge, bound charge, and free charge densities
$\rho\boldsymbol{\Gamma}$	Torque density
$\mathbf{r}_i$	Position of molecule i
$\mathbf{r}$	Position vector $\mathbf{r} = (x, y, z)$
R	Tube radius
$\mathbf{R}$	Dipole flux tensor
$\mathbf{R}_\alpha$	Atom α position vector with respect to c.o.m.

σ	Lennard–Jones length scale, or production term, or stress tensor component, or electric conductivity
σ_{path}	Characteristic mean 'interaction-free' length scale
σ_S	Width of Stern layer
Σ_W	Wall charge
S	Dynamic structure factor
S_{bb}	Static polarization autocorrelation function
$\mathbf{S}_i$	Spin angular momentum of molecule i
S_e	Seebeck coefficient
S_T	Soret coefficient
$\Theta, \Theta_{\mathrm{mol}}, \Theta_P, \Theta$	Moment of inertia per unit mass, molecular moment of inertia, principal molecular moment of inertia, and average of Θ_P
$T, T_{\mathrm{cl}}, T_{ext}$	Kinetic temperatures
U	Characteristic velocity, or potential function
u_{EO}	The Helmholtz–Smoluchowski velocity
$\mathcal{V}$	Fluid volume element
$\mathbf{v}_i$	Velocity of molecule i
w	Slit-pore width
W_{rz}	Reaction zone width of chemical wavefront
Wo	Womersley number
$\mathbb{X}$	General driving force
x_A	Mass fraction of molecule A
ζ	Abbreviation for either ζ_l or ζ_t, or electric potential at Stern-diffuse layer
$\zeta_0, \zeta_v, \zeta_r$	Shear, bulk, rotational spin viscosities
ζ_l	Longitudinal spin viscosity
ζ_N	Navier friction coefficient
ζ_t	Transverse spin viscosity $\zeta_t = \zeta_0 + \zeta_r$

1   Introduction

Hydrodynamics describes fluid dynamical properties like mass density, streaming velocity and energy [15]. Understanding and controlling fluids has a long history [50], and despite the prefix, hydrodynamics extends today beyond the study of water, and more generally to all fluids. The definition of a fluid is ambiguous, but we here think of it in an intuitive manner as a material (or substance) that will easily flow when a force is applied to it. Liquids and gasses are two usual examples of a fluid.

The fundamental assumption in hydrodynamics is that the properties vary sufficiently smoothly in both time and space; this is known as the continuum hypothesis, see, for example, Ref. [141]. In this way the properties can be treated mathematically as field variables. The hypothesis is in agreement with our everyday experience: when we are stirring a cup of coffee or riding a bicycle, the fluid flow appears smooth. However, the continuum hypothesis is not strictly true. Think of a small fluid volume (or fluid element), denoted V, embedded in a material at rest. Hydrodynamics will predict that the mass is constant with respect to time; however, due to thermal motion, molecules will enter and leave the fluid element, and the mass of V will fluctuate. As the element volume increases or if we perform a sufficiently long time average, the mass of V will converge to that of the hydrodynamic prediction. What defines sufficient large volumes and time averages is not clear, and we return to this example later in this chapter.

Channels and tubes for fluid flows with nanoscale cross section and nanoscale volume can now be fabricated with impressive accuracy [88, 170]. In order to control and utilise such nanoscale fluid volume devices, it is important to develop models that can describe and predict the fluid dynamics. Is the continuum picture, which has been applied with great success to micro- and macro-fluid systems, also applicable on the nanoscale? From the preceding example, it appears that the answer is not trivially 'yes' or 'no', and this question is the underlying theme of the text.

We must settle on a few important definitions. Eikel and van den Berg [61] define *nanofluidics* as the study and application of fluid flows confined in and around nanosized structures. For fluids confined in nanoscale structures (or geometries) a non-negligible fraction of the fluid molecules will interact with the wall atoms. To describe such systems in detail an in-depth knowledge of both the wall–fluid and fluid–fluid interactions are needed [29, 117, 165]. The complexity can be overwhelming and appear almost intractable from a modelling point of view. Fortunately, many of the underlying physical mechanisms relevant in confined fluids are also present in the non-confined case, and one can simplify the problem considerably by studying these systems

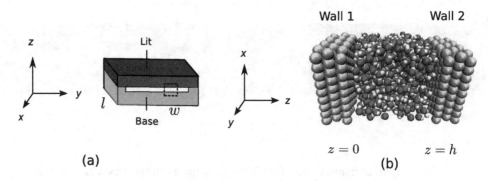

Figure 1.1 (a) Illustration of a slit-pore geometry fabricated by etching a silicon base and placing a lit on the cavity. (b) A snapshot of a computer simulation of water flowing in a region of the slit-pore, highlighted by the punctured box in (a). In the simulations, the walls and fluid are infinite in extent in the (x, y)-plane. The pore height h is 3.4–3.5 nm. From Ref. [102] with permission.

[49, 66, 110, 199]. One can even study some of the mechanisms for non-confined systems in equilibrium [31, 90], that is, in no-flow situations. Therefore, in order to explore nanofluidic systems in detail, we here extend the scope of the book and define *nanoscale hydrodynamics* as

> *the study and application of fluid systems, where the system characteristic length scale is in order of nanometres.*

Whether the system is confined or not, a *nanoscale fluid system* is then a system where the characteristic length scale is in order of nanometres, that is, 1–100 nanometres. Naturally, the length scale needs to be clearly defined for each system. Notice that the related field *nanofluids* is the study of nanosized particles suspended in fluids and is not the focus of the text.

An example of a nanoscale fluid system that we will explore is water confined in a slit-pore, where the pore height, h, is around 10 water diameters or 3.4–3.5 nm. In brief, the slit-pore geometry can be fabricated experimentally from etching a base (typically a silicon wafer) and placing a lit on the formed cavity; see Fig. 1.1 (a). Figure 1.1 (b) is a snapshot from a molecular dynamics, MD, simulation that simulates only a small region of the slit-pore and after the coordinate system has been rotated twice. We have that $w \gg h$ and $l \gg h$, and h is the characteristic length.

We will often use this simple geometry in our study of confined nanoscale fluids. Compared to clever choices of coordinate system, where, for example, the walls are located at dimensionless points $z = \pm 1$, the coordinate system shown in Fig. 1.1 (b) leads to slightly more complicated mathematical expressions. However, the results will depend explicitly on h, the characteristic length scale, and we will stick with this more intuitive choice of coordinate system with the exception of Section 5.6, where we explore molecular fluid flows.

In confinement one can decompose the forces acting in the system into surface forces (e.g., the wall-fluid frictional force) and volume forces (e.g., the gravitational force). The total surface force and volume force must be proportional to the wall–fluid surface area and fluid volume, respectively. For the geometry in Fig. 1.1 we have

$$\frac{\text{Total surface force}}{\text{Total volume force}} \propto \frac{l \times w + h \times l}{w \times l \times h} \approx \frac{1}{h}, \tag{1.1}$$

as $h \ll w$. This is also known as the square-cube law [1] and shows that the surface force becomes dominant for small characteristic lengths. One immediate result of this is that confined nanoscale fluid flows cannot be generated by, say, Earth's gravitational pull, but must be realised through application of more advanced methods such as electro-osmosis.

In nanofluidic laboratory experiments the fluid velocity rarely exceeds 0.1 m/s; see Whitby and Quirke [212], corresponding to a flow rate less than 10^{-15} L/s. If we define the Reynolds number, Re, as

$$\text{Re} = \rho h U / \eta_0, \tag{1.2}$$

where ρ is the mass density, $U = 0.1$ m/s is the characteristic fluid velocity, and η_0 is the shear viscosity, the Reynolds number is in practise below 0.01. The flow is therefore a Stokes flow, or creeping flow, and we can safely neglect the advective inertial forces in the hydrodynamic description of the system.

Molecular dynamics is widely used today to study nanoscale fluid systems. Common for these computer-based studies is that unrealistically large flow velocities are simulated. Usually the velocity is in order of 10–10^2 m/s, nevertheless, due to the small length scales, the Reynolds number is still small, often between 1 and 10, and inertial forces can again be ignored when analysing the simulation data.

Another important point about simulation studies is that despite the very large fluid velocities, the flow speed is usually significantly lower than the corresponding sound speed, c_s. This is commonly quantified through the Mach number,

$$\text{Ma} = U / c_s. \tag{1.3}$$

If Ma < 0.3, the fluid compressibility effects can usually be ignored. For water at ambient conditions, c_s is on the order of 10^3 m/s and the corresponding Mach number is below 0.1 in simulations under usual simulation conditions. Nanoscale flow systems, both real and simulated, are thus characterised as being laminar and incompressible, and this simplifies the mathematical analysis of the models considerably, as we will see.

In our exploration of nanoscale hydrodynamics we focus on fluid systems characterised by a system relaxation time that is sufficiently small compared to the time scale at which we perform measurements or simulations. Specifically, if τ_s is the time for the fluid internal stress to relax after shearing/deformation and τ_{obs} is the time we observe the system, we define the Deborah number

$$\text{De} = \tau_s / \tau_{\text{obs}}, \tag{1.4}$$

which then must be significantly smaller than 1. For fluids like methane, butane, and water at ambient conditions, a small Deborah number is easily obtained, even in computer simulations. On the other hand, polymer and glass systems feature very large relaxation times having a large Deborah number even if τ_{obs} is the real laboratory observation time.

We here refer to fluids with sufficiently small τ_s compared to τ_{obs} as simple fluids; τ_{obs} is often defined by what can be achieved in computer simulations. By simple systems we mean simple fluids, either unconfined or confined to simple straight channels like slit-pores or nanochannels.

Nanofluidics is believed to play a critical role in many areas of future engineering [61, 132, 176]. The purpose here is not to present the many exciting applications, but to show how nanoscale fluid systems can be modelled and what new insight into fluid and liquid theory this brings. Often what appears to be a new phenomenon specific to the nanoscale is actually omnipresent, but can be ignored on larger length scales. The term *nanoscale fluid phenomenon* is still used when the phenomenon is particularly relevant to nanoscale fluid systems.

1.1 Nanoscale Fluid Phenomena

Before going into detail on how to model nanoscale fluid systems and what we can learn from that, it will be enlightening to see a few examples of some of the phenomena we will explore later. The examples will by no means cover all phenomenology, but they illustrate at least some unique features of these systems and also motivate the topics in the book.

1.1.1 Flows in Nanochannels

Horn and Isrealachvili [114] showed in 1981 that the force acting between two mica surfaces in liquid octamethyl-cyclotetra-siloxane (OMCTS) features oscillations as the distance between the mica surfaces is varied. Specifically, the oscillations have a period matching the diameter of OMCTS, and the amplitude decay as the distance between the surfaces is increased. This seminal result is a fingerprint of a molecular layering near the surface, a layering which is strong in the fluid region close to the surface and decays with distance. This layering was also reported in 1977 by Toxvaerd and Præst-gaard [202] who studied confined systems using molecular dynamics simulations. The focus here is on the hydrodynamics, and we therefore ask how the layering affects the fluid flow properties. For simple fluids, as we have defined it above, the effect is surprisingly small, but for non-simple fluids composed of, say, long alkane chains the picture is much more complicated, and the fluid's flow resistance increases significantly as we reach nanoscale confinement [83]. This increase is attributed to adhesion/cohesion effects coming from molecular layering, but also to crystallisation, vitrification, phase transitions, and more; see a summary in Ref. [68]. The increased flow resistance has led to the concept of an effective viscosity, η_{eff}, which in nanoscale systems can be many times larger than the shear viscosity η_0 characterising the flow properties in the macroscopic or non-confined case.

The opposite effect due to confinement, namely a flow enhancement, has also been observed. This phenomenon can be quantified through the enhancement coefficient, E, which is defined as the ratio between the experimentally measured volumetric flow rate, Q_{exp}, and the theoretical predicted flow rate, Q_{the},

$$E = Q_{exp}/Q_{the}. \tag{1.5}$$

Recall, the volumetric flow rate is the fluid volume discharged by the channel per time unit [38]. The predicted volumetric flow rate Q_{the} is often calculated from the Navier–Stokes equation in the given geometry, using η_0, and with specified boundary conditions; thus, we assume that the Navier–Stokes equation is applicable on the nanoscale.

In 2005, Majumder et al. [153] investigated water flow through carbon nanotubes embedded in a membrane and having diameters of around 7 nm. Using zero velocity boundary conditions to calculate Q_{the}, the authors reported a surprisingly large enhancement coefficient, on the order of 10^4. One hypothesis for the mechanism behind the enhancement is that the fluid velocity at the wall–fluid boundary is non-zero. This is referred to as slippage and can be quantified from the slip enhancement coefficient,

$$E^{slip} = Q_{the}^{slip}/Q_{the}^{noslip}, \tag{1.6}$$

where Q_{the}^{slip} is the theoretical prediction for the volumetric flow rate using slip boundary conditions and Q_{the}^{noslip} is the prediction using the traditional no-slip boundary conditions. For a steady flow in slit-pore geometries (Fig. 1.1), the slip-enhancement coefficient can be evaluated to

$$E^{slip} = 1 + 6L_s/h, \tag{1.7}$$

where L_s is the slip length. The slip length is the interesting quantity when discussing flow enhancement, and is, for the planar Poiseuille flow, the distance away from the wall–fluid interface to where the linearly extrapolated velocity is zero. The slip length is illustrated in Fig. 1.2 (a).

For highly hydrophobic walls, the slip length is expected to be large; the results for carbon nanotubes from Majumder et al. suggest L_s to be on the order of 10^4 nm. The actual slip length magnitude is still debated. For example, from molecular dynamics simulations Kannam et al. [125] found a slip length of around 100 nm in a carbon nanotube with a diameter of 4 nm; this corresponds to a plug-like flow. From experiments on a Landau–Squire flow, Secchi et al. [188] found the same slip length, but for carbon nanotubes with diameters of around 50 nm. There is general agreement that the slip length decreases as the tube diameter increases and that it converges to that of graphene; experimental, theoretical, and simulation studies find $L_s = 10$–80 nm for graphene–water; see Ref. [125] and references therein. Figure 1.2(b) plots the slip enhancement coefficient for graphene–water as a function of slit-pore height h.

Importantly, fluid slippage also occurs on macroscopic length scales, but from Eq. (1.7) the effect of this on the flow rate is not observed under usual macroscopic

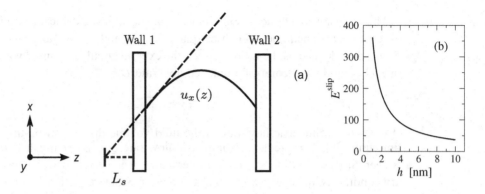

Figure 1.2 (a) Illustration of the slip length, L_s, for the planar Poiseuille flow. The dashed line is the tangent line for the fluid velocity at $z = 0$. The curve illustrates the fluid velocity x-component. (b) Slip enhancement coefficient for a slit-pore where $L_s = 60$ nm (graphene–water).

circumstances, as h is many orders of magnitude larger than the slip length. Even for microscopic length scales, E^{slip} is close to unity.

1.1.2 Capillary Raise

A common way to fill nanochannels and nanopores with a fluid is through capillary filling and, naturally, this method has drawn a lot of attention to the research community. Capillary filling in micro- and macropores is, under usual conditions, described satisfactory by the Lucas–Washburn equation, however, on the nanoscale the slippage phenomenon introduced in the previous section becomes important. The modified Lucas–Washburn equation including this effect reads [57, 119]

$$h^2_{\mathrm{cap}}(t) = \frac{\gamma R \cos(\theta)}{2\eta_0}\left(1 + \frac{4L_s}{R}\right)t, \tag{1.8}$$

where h_{cap} is the capillary height, γ is the surface tension, R is the radius of the tube, and θ is the contact angle between the meniscus and the wall; see Fig. 1.3 (a).

Figure 1.3(b) shows the capillary height h_{cap} as a function of time for two different model fluids investigated using molecular dynamics (MD) simulations. It is worth noting that the typical unit length scale in molecular dynamics is 3–5 Å and the unit time scale is on the order of picoseconds. After a short inlet transient time, the capillary height predicted by the modified Lucas–Washburn equation is confirmed – even quantitatively.

The research continues, as questions remain unresolved. For example, it is known that the contact angle and the slip length are correlated quantities [211], and this calls for a revision of the fundamental theory behind capillary filling on the nanoscale.

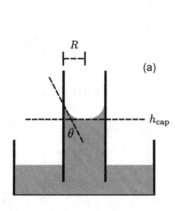
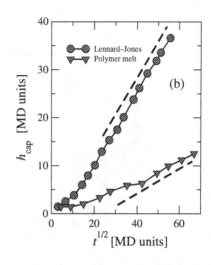

(a) Illustration of capillary raise. (b) Capillary height as a function of time in a tube of radius 10 (in MD units); filled circles are data for a Lennard–Jones liquid and triangles data for a model polymer melt. Punctured lines illustrate the theoretical predictions for the slopes. From Dimitrov et al. [57].

1.1.3 Anisotropy in Confined Dielectrics

Application of an external electrical field, $\mathbf{E}^{\text{ext}}$, to a dielectric material gives rise to a polarisation, $\mathbf{P}$. Recall, in the static, homogeneous, isotropic, and linear cases, the polarisation is

$$\mathbf{P} = \varepsilon_0(\varepsilon_r - 1)\mathbf{E}^{\text{ext}} = \varepsilon_0 \chi_e \mathbf{E}^{\text{ext}}, \tag{1.9}$$

where ε_r and ε_0 are the relative and vacuum dielectric permittivities, respectively, and $\chi_e = \varepsilon_r - 1$ is the electric susceptibility. In the situation where an external electric field is suddenly switched on at, say, $t = 0$, the system relaxation response can also be studied. The simplest model for this non-static case is the Debye model,

$$\mathbf{P}(t) = \varepsilon_0 \chi_e \mathbf{E}^{\text{ext}}(1 - e^{-t/\tau_D}); \tag{1.10}$$

τ_D is the Debye relaxation time. Notice that the polarisation $\mathbf{P}$ converges to $\varepsilon_0 \chi_e \mathbf{E}^{\text{ext}}$ in accordance with the static case, Eq. (1.9).

Figure 1.4 shows molecular dynamics results for the polarisation as a function of time for water confined in a slit-pore. As the field is applied parallel to the walls, the system response is bulk-like and follows an exponential relaxation in accordance with the Debye model. On the other hand, the response is significantly changed when the field is applied normal to the walls; here it resembles a small amplitude step response. In both experiments [76] and simulations [149] this anisotropy is found for slit-pore heights up to 100 nm, and we approach the microfluidic length scale.

The reduced polarisation phenomenon indicates that there exist a parallel permittivity $\varepsilon_r^{\parallel}$ and a normal permittivity $\varepsilon_r^{\perp}$ with respect to the wall plane. If the system's dielectric response is isotropic, we have $\varepsilon_r^{\parallel} = \varepsilon_r^{\perp} = \varepsilon_r$.

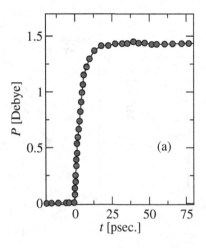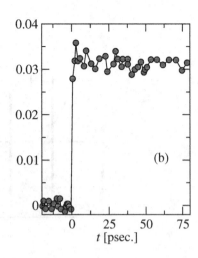

Molecular dynamics results for the polarisation as a function of time for water confined in a slit-pore with $h = 11$ nm; see the geometry in Fig. 1.1 (b). (a) The external electric field is applied parallel to the wall at $t = 0$. Same as (a), but where the field is applied normal to the wall. Data are from Ref. [149].

To model the reduced normal permittivity, we can divide the confined fluid into layers with respect to the z-direction. These layers can be considered as capacitors in a serial arrangement. Each capacitor has capacitance C_i, permittivity $\varepsilon_{r,i}^{\perp}$, and separation (or height) h_i. The fundamental idea now is that the capacitance in the wall–fluid interface is low, as the water density is low. The total capacitance is

$$\frac{1}{C} = \sum_i \frac{1}{C_i} . \tag{1.11}$$

The capacitance for each layer is $C_i = \varepsilon_{r,i}^{\perp} \varepsilon_0 A / h_i$, where A is the surface area of the capacitor, and we get

$$\varepsilon_r^{\perp} = \frac{h}{\sum_i h_i / \varepsilon_{r,i}^{\perp}} . \tag{1.12}$$

Zhang [215] studied the simplest scenario of a fluid next to a wall and included only two fluid layers: one layer (denoted layer 1) just adjacent to the wall, where the height was h_1, and a layer 2 with height h_2, such that $h = h_1 + h_2$. If h_1 is on the order of angstroms, then $\varepsilon_{r,1}^{\perp} \approx 1$, as this interfacial region is almost a vacuum. The second layer, we assume, has permittivity $\varepsilon_{r,2}^{\perp} \approx \varepsilon_r$. Then Eq. (1.12) simplifies to $\varepsilon_r^{\perp} = h/(h_1 + h_2/\varepsilon_r)$. As $h \to h_1$ we have $h_2 \to 0$ and $\varepsilon_r^{\perp} \to 1$. As h increases we reach the regime where $h_2 \gg h_1$ and $h_1 + h_2/\varepsilon_r \approx h_2/\varepsilon_r$, that is, $\varepsilon_r^{\perp} \approx \varepsilon_r$. Thus, this simple capacitor model predicts the monotonic increase in the permittivity as a function of h. Note, however, that Ballenegger and Hansen have questioned the layering picture [10].

As the dielectric response is anisotropic, the permittivity cannot be described by a single scalar, but must be considered as a general tensor property [30], in this case a so-called rank-2 tensor which will be introduced in Chapter 2. For some dielectric materials, for example, multi-component crystals, the anisotropy can result in a polar-isation which is not parallel to the electric field. To complicate the problem further, the preceding model also indicates that the permittivity is position dependent, that is, the permittivity tensor is a function of position in general.

It is not only the dielectric response which features anisotropy and position depend-ency [102]. Mechanical properties like viscosity also should, in principle, be considered to be anisotropic and position dependent. In our exploration we will see that this is complication is usually not needed for simple systems unless we study the dielectric properties.

1.1.4 Coupling Phenomena

In a molecular dynamics study, de Luca et al. [150] investigated water confined in a nanoscale slit-pore, where a rotational electric field was applied to the system; see the illustration in Fig. 1.5. The authors designed the slit-pore such that one wall was a graphene wall, a hydrophobic material, and the other wall was β-cristobalite which is hydrophilic. The water molecules' dipoles will align with the field; of course, due to thermal fluctuations this alignment is far from perfect. As the field rotates, the molecules will also rotate because the field exerts a torque on the dipoles and in this way the water molecules obtain a non-zero average angular momentum. Due to con-servation of total angular momentum, this intrinsic molecular rotation results in a fluid flow, that is, a fluid translational motion. This coupling is not included in the classical hydrodynamic description, where the local rotation is given directly by the (local) curl of the streaming velocity field and is therefore not treated as an independent dynamical variable.

On a small historical note, the coupling was already described in the late 1890s by the Cosserat brothers [43, 44] and again treated in great detail in the 1950s to the 1980s [3, 52, 67, 190]. With the increasing interest in nanoscale hydrodynamics in the 2000s it is again the focus of many research groups [28, 72, 150].

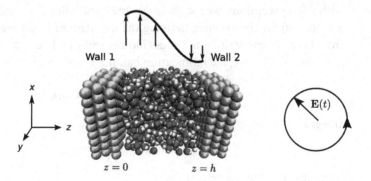

Figure 1.5 Illustration of the resulting velocity profile when a rotating electric field is applied to a slit-pore with water and non-symmetric wall hydrophobicity. Here wall 1 is hydrophobic, giving large slippage, and wall 2 is hydrophilic, giving small slippage. Figure from Ref. [102] with permission.

This particular coupling phenomenon has an interesting potential application. Pumping fluids that are confined in nanoscale geometries is a challenging task. Naively, one can apply a pressure difference, Δp, over the channel inlet and outlet. In a first approximation, assume that the volumetric flow rate is proportional to the pressure difference applied, $Q = \Delta p/R_{\text{hyd}}$, where R_{hyd} is the hydraulic resistance. It can be shown from classical hydrodynamics that $R_{\text{hyd}} \propto 1/h^4$ [38], and since h is in the order of nanometres, the hydraulic resistance is extremely large, leaving this simple pumping device unusable under normal circumstances. Other pumping mechanisms are therefore needed, and exploiting the coupling between the molecular rotation and the translational motion is one possibility. Also, Felderhof [71] proposes nano-propulsion systems based on this coupling.

The coupling between the molecular rotation and the flow is just one example of many coupling phenomena [52] relevant in nanoscale hydrodynamics. In nanoscale systems very large thermal gradients can be achieved, and these gradients can result in different mass fluxes for fluid mixtures, an effect referred to as the Soret effect. Also, Bresme et al. [35] showed that for polar fluids, a large thermal gradient induces polarisation. We return to these coupling phenomena in Chapter 6.

1.1.5 Non-local Viscous Response

The final example is from Todd et al. [199]. Here the authors investigated the fluid shear stress when applying a sinusoidal shear force to the fluid. The shear force acts in the x-direction and is a function of the z-direction; here we write it in terms of a force density $\rho g = \rho g_0 \cos(kz)$, where g_0 is the acceleration amplitude, and k defines wavelength of the imposed force; see Fig. 1.6. We assume that the system is homogeneous and infinite in extent, so we need not consider effects from confining walls. For this system there is only one non-zero shear stress component, σ_{zx}, where the first index indicates that the normal vector to the sheared virtual fluid surface is parallel to the z-direction, and the second index indicates the force direction; see Fig. 1.6. We here follow the original work and discuss the system response to the shear force through the stress; however, with a few exceptions, we use the shear pressure rather than the shear stress, as this is a more natural choice when deriving the momentum balance equation. Also, we will from here on omit writing the stress indices and simply use $\sigma = \sigma_{zx}$.

In the steady state, the momentum balance equation reads

$$\frac{\partial \sigma}{\partial z} = -\rho g = -\rho g_0 \cos(kz). \tag{1.13}$$

Integration gives

$$\sigma(z) = -\frac{\rho g_0}{k} \sin(kz), \tag{1.14}$$

using that the stress is zero at $z = 0$.

The stress can also be predicted using Newton's law of viscosity. For the geometry here we have

$$\sigma(z) = -2\eta_0 \dot{\gamma}(z), \tag{1.15}$$

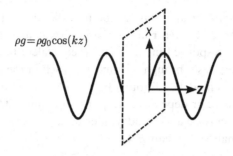

$$\rho g = \rho g_0 \cos(kz)$$

Illustration of the applied sinusoidal shear force and the fluid shear surface. The fluid is not shown.

where $\dot{\gamma}$ is the strain rate. For the applied force used here, the strain rate is a sine function, $\dot{\gamma} = \tilde{\gamma}(k)\sin(kz)/2$, hence,

$$\sigma(z) = -\eta_0 \tilde{\gamma}(k)\sin(kz). \qquad (1.16)$$

Both η_0 and the strain rate amplitude $\tilde{\gamma}$ can be found from independent methods.

The shear stress obtained from the momentum balance equation, Eq. (1.14), and the shear stress resulting from Newton's viscosity law, Eq. (1.16), are both shown in Fig. 1.7. The z-coordinate is given in units of around one atomic diameter, 3–4 Å. For small k-values (long wavelengths) Newton's viscosity law predicts the shear stress satisfactorily; however, as the wavelength approaches the atomic length scale (large k-values), the stress features significant reduction compared to the prediction from Eq. (1.16). It is worth noting that the force is sufficiently low such that the system is in the linear response regime [94].

The mechanism for this stress reduction is believed to reside in the non-local nature of the fluid response to the shearing force. Newton's viscosity law is local in the sense that the stress at some point, $\mathbf{r}$, is proportional to the strain rate at that given point. More generally, the stress at $\mathbf{r}$ is dependent on the entire system strain rate distribution.

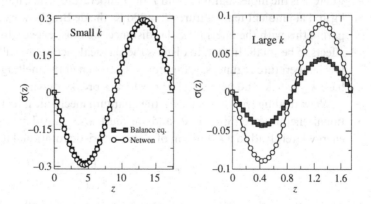

Shear stress profiles for an atomic fluid subjected to a sinusoidal shear force. In the small k-value figure the wavelength is one order of magnitude larger than in the large k-value figure. The z-coordinate is given in units of approximately one atomic diameter and the stress is in computer simulation units. Data redrawn from Ref. [102].

Phenomenologically, this can be modelled by letting the viscosity be a function of the distance between $\mathbf{r}$ and all other points in the system. That is, the viscosity is a non-local response function. This is equivalent to how we model the temporal visco-elastic response through a memory kernel; in the most general formulation the fluid response is modelled through time- and space-dependent, or equivalently frequency- and wavevector-dependent, response functions. This generalisation was initiated by Boltzmann [174] and is today referred to as generalised hydrodynamics; we return to this formalism in Chapter 4.

1.2 Nanoscale Hydrodynamic Modelling

How can we then model, that is better understand, these nanoscale hydrodynamic phenomena? The preceding discussions have all been based on hydrodynamic theory, that is, the continuum picture. However, we should be a bit concerned for at least two reasons. (i) One would think that the intrinsic discrete nature of the fluid at very small scales will destroy the continuum picture. This is also pointed out by Lautrup, who categorises continuum modelling as physics on the macroscopic scale [141]. (ii) At these small length scales the system may not behave according to classical mechanics, but be quantum mechanical in nature.

Let us first address the latter. Quantum mechanical effects become relevant when the system characteristic length scale is on the order of the de Broglie wavelength, λ_{Br}. The characteristic length scale is not always well defined; here we use the average distance between the molecules' centre of mass [90], which for liquids composed of small and approximately spherical molecules is around 3–4 Å. The de Broglie wavelength is [87]

$$\lambda_{\mathrm{Br}} \approx \frac{1}{\sqrt{mT}} 10^{-22} \mathrm{m}\sqrt{\mathrm{kg\ K}}, \tag{1.17}$$

where m is the molecular mass and T the temperature. Thus, for water $m = 2.99 \times 10^{-26}$ kg and at ambient temperatures $T = 300$ K, the de Broglie wavelength is 0.33 Å. Comparing this with the intermolecular distance, we must expect the quantum mechanical effects to be small. In fact, to dismiss the classical picture we will need to be in the very low temperature regime; say, for atomic hydrogen at the melting point $T = 13.95$ K we have $\lambda_{\mathrm{Br}} \approx 6$ Å, and here classical mechanics breaks down.

When dealing with molecules in the quantum mechanical realm, the molecular rotational energies are discretised, and we must also consider this effect. The different energy levels lead to a definition of a characteristic rotational temperature,

$$T_{\mathrm{rot}} \approx \frac{4 \times 10^{-46}\mathrm{kg\ m^2\ K}}{\Theta_{\mathrm{mol}}}, \tag{1.18}$$

where Θ_{mol} is the molecular moment of inertia. For small molecules, Θ_{mol} is on the order of $10^{-47} - 10^{-45}$ kg m^2 and the characteristic temperature typically fulfils $T_{\mathrm{rot}} <$ 50 K. Therefore, for the rotation we require that $T \gg T_{\mathrm{rot}}$ if quantum mechanical effects are ignored, and this is the case for the systems we explore here. We will not treat molecular vibrational degrees of freedom in this text, and therefore assume that this set of dynamics has no effect on the phenomena we investigate.

What about molecular discreteness? In the continuum model we picture the fluid as being composed of small fluid elements. Each such element contains a sufficiently large amount of molecules such that one can define the same quantities (or properties) for the local fluid element as those of the fluid itself, no matter how small the fluid element is [207]. The quantities are here denoted hydrodynamic variables, or hydrodynamic quantities, and can be mass density, streaming velocity, energy, and so on.

Returning to the example of the fluid element $\mathcal{V}$ at the beginning of this chapter, we follow Lautrup [141] and consider the mass density, $\rho_{\mathcal{V}} = \rho_{\mathcal{V}}(t)$. The mass of $\mathcal{V}$ is simply the total mass of the constituent molecules inside $\mathcal{V}$, and $\rho_{\mathcal{V}}$ can then intuitively be written as

$$\rho_{\mathcal{V}}(t) = \frac{1}{\Delta\mathcal{V}} \sum_{i\,\text{in}\,\mathcal{V}} m_i \,, \tag{1.19}$$

where index i runs over all molecules contained in $\mathcal{V}$, m_i is the mass of molecule i, and $\Delta\mathcal{V}$ is the fixed fluid element volume. In the absence of a flow, $\rho_{\mathcal{V}}$ will fluctuate in time around the average density, ρ_{av}, as molecules enter and leave the fluid element due to thermal motion. The classical continuum model predicts a time-independent density $\rho_{\mathcal{V}}(t) = \rho_{av}$, or equivalently a time-independent number of particles N_{av} in $\mathcal{V}$. The fluctuations can then be thought of as a measure of the deviation between the continuum model and the actual situation.

Let us then quantify the fluctuations by the standard deviation, σ, around the average. If the molecule entering and leaving the fluid element is a true random event, then the standard deviation is proportional to $\sqrt{N_{av}}$. It is not the absolute standard deviation itself which is interesting here; one naturally expects the deviation to be smaller in the dilute case, as fewer molecules enter and leave the fluid element. Instead one can study the relative standard deviation σ/ρ_{av} which is proportional to σ/N_{av} or $1/\sqrt{N_{av}}$ when the volume $\Delta\mathcal{V}$ is fixed. Therefore, if we accept a relative error in the order of 1 per cent, we have that the average number of molecules in the fluid element must be greater than 10^4. There is no general rule for the acceptance threshold value; and this is perhaps not the interesting point here. The interesting point is that the relative standard error decreases with increasing density as $\sigma/\rho_{av} \propto 1/\sqrt{\rho_{av}}$. Thus, the continuum model performs poorly at small length scale for gasses when compared to more dense systems at the same length scale; also see Ref. [127].

To elaborate further, we study another very important hydrodynamic variable, namely the streaming velocity. This time we do not make the simple statistical arguments we have just made, but rely on a molecular dynamics simulation that includes correlation effects and so on; the molecular dynamics technique is introduced in what follows. In the first-order approximation we can ignore the density fluctuations [102], and the streaming velocity, $\mathbf{u}_{\mathcal{V}} = \mathbf{u}_{\mathcal{V}}(t)$, of $\mathcal{V}$ is intuitively given by

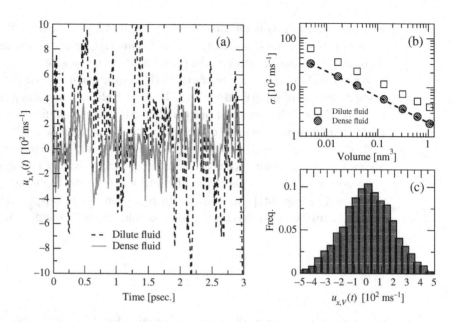

(a) Streaming velocity of a cubic fluid element with a volume of approximately 1 nm^3. (b) The corresponding standard deviation. (c) The normalised histogram for the streaming velocity x-component. The densities of the methane fluids are $\rho = 270\,\text{kgm}^{-3}$ and $\rho = 540\,\text{kgm}^{-3}$, and the temperature is $T = 222\,\text{K}$.

$$\mathbf{u}_{\mathcal{V}}(t) = \frac{1}{\rho_{\text{av}}\Delta\mathcal{V}} \sum_{i\,\text{in}\,\mathcal{V}} m_i \mathbf{v}_i(t) \,, \qquad (1.20)$$

where $\mathbf{v}_i$ is the velocity of molecule i. Note that the streaming velocity is defined from the molecular momenta and gives the correct mass weighted average. Again, there is no advection in the system, and we simply study the effect of thermal fluctuations. Figure 1.8(a) shows molecular dynamics simulation results for the x-component of the streaming velocity of a cubic fluid element with $\Delta\mathcal{V} \approx 1$ nm^3 for two cases: (i) a relative dilute methane fluid and (ii) a relative dense methane fluid. One can see that the fluctuations in the dilute case are larger than in the dense case. The underlying reason for this important result is that in the dense case the molecules collide, or more precisely interact, very frequently resulting in a high degree of momentum exchange compared to the dilute case. The statistics are summarised in Fig. 1.8(b), where the standard deviations around zero for the two state points are shown for different fluid element volumes. The punctured line shows a power-law function with an expected exponent of $-1/2$. If we accept a standard deviation threshold of around 200 ms^{-1}, that is, we demand 68 per cent of the data points to be within $u_{x,\mathcal{V}}(t) = (0\pm2) \times 10^2$ ms^{-1}, then the continuum model fails for characteristic length scales below 1 nm in the dense situation. This length scale increases as the density decreases in agreement with the preceding discussion of momentum exchange. In Fig. 1.8(c) the normalised frequency for $u_{x,\mathcal{V}}$ is plotted in the dense case; this plot indicates that the fluctuations are Gaussian distributed.

Strictly, the definitions of mass density and streaming velocity, Eqs. (1.19) and (1.20), make no sense in the coarse-grained continuum picture since there are no such things as molecules. They must therefore be thought of as microscopic (or molecular) definitions of the hydrodynamic variables, and not as continuum definitions.

If we insist on a *classical* continuum picture, the thermal fluctuations are indeed problematic. However, we could choose to include the fluctuations in our modelling and, inspired by the Langevin equation, one strategy that comes to mind is to simply add a stochastic noise (or force) term to the dynamical equations for the hydrodynamic variables. In this way the many degrees of freedom behind the thermal fluctuations are coarsened into a single random term, keeping the problem low-dimensional. Adding a fluctuating stochastic term onto the dynamics means adding energy to the system, and this must be balanced correctly by an energy drainage, or dissipation, originating from the system transport processes. Thus, the stochastic force amplitude cannot be chosen ad hoc, but depends on the system transport coefficients and the temperature; this is the so-called fluctuation-dissipation theorem. We will adopt the stochastic force method; however, we only explore situations where we do not need to invoke the fluctuation-dissipation theorem, and simply require that the force has the following properties:

1. the average over an ensemble is zero, and
2. it is uncorrelated with respect to the hydrodynamic variables.

The term 'ensemble' is here used in the general sense as a 'sufficiently large and statistically independent set' and not as a specific statistical mechanical ensemble. Adding such a stochastic force term to the dynamical equations for the hydrodynamics variables leads to a set of stochastic differential equations which pose new challenges. However, by performing an ensemble average over a set of independent initial conditions, the equations become deterministic due to the stochastic force properties we have just listed. For a more careful treatment and discussion of how to treat these hydrodynamic fluctuations, the reader is referred to the original work by Landau and Lifshift [139, 140] as well as the book by Zárate and Sengers [56].

By averaging we then suppress the thermal fluctuations, but this still does not answer our original question, namely whether the continuum picture can be applied to nanoscale fluid systems, or more precisely, whether the underlying physico-chemical processes can be modelled using continuum theory; this may fail even in the absence of fluctuations. We shall address this fundamental question in great detail throughout the book; but before doing so, we saw in the preceding examples one important point which is worth recalling:

In the continuum picture, the dynamics of the different hydrodynamic variables, say, the mass and momentum densities, are given through the balance equations. The balance equation is a partial differential equation; however, as such it does not form a mathematical closed problem that can be solved. Therefore, one applies constitutive equations (or relations) that typically relate the diffusive processes with the system gradients. The constitutive relations are models, and not fundamental laws of physics. Newton's law of viscosity, which we discussed for the shear stress, is an example of such a model. Then, the continuum description is based on the balance equation *and* a set of constitutive relations. If the theory fails, it may not be because the continuum picture fails, but because the constitutive relations do not model the underlying physico-chemical processes appropriately. The non-local shear response is an example where the classical constitutive relation is not appropriate; however, by proposing a generalised constitutive relation the continuum picture is indeed applicable.

Thus, what may appear to be a breakdown of the continuum picture can simply be a result of poor modelling and lack of generality.

Often the next step in the theoretical treatment of nanoscale hydrodynamics is based on Mori–Zwanzig projection-operator formalism [161, 219] that leads to a generalised Langevin-type equation for the hydrodynamic variable. This equation depends on a function describing the transport properties, and this can be found from simulations or theoretically from mode coupling theory [81]. A related yet slightly different theory is the generalised collective mode theory [54], where the set of hydrodynamic variables is increased, leading to a larger dynamical space and consequently a better quantitative agreement between data and theory [40]. These advanced theories are successful in predicting important phenomenology; however, these are not the topic here, and while we will treat a lot of the same phenomena, we will do so from a purely hydrodynamic viewpoint.

1.3 Molecular Dynamics

The microscopic definitions of the hydrodynamic variables given in Eqs. (1.19) and (1.20) are based on the molecular positions and momenta. If a system is composed of N molecules, we can envision that a particular system state is defined by $3N$ position coordinates and $3N$ momentum coordinates; this is the molecular phase space point. As the molecular positions and momenta evolve in time, the phase space point changes, resulting in the system phase space trajectory. Molecular dynamics (MD) is a powerful simulation method to trace out this phase space, or at least parts of it and hopefully the important parts. From the dynamics of the phase space we can gain general knowledge of the fluid properties and in particular the hydrodynamics.

It is worth mentioning that many other microscopic and mesoscopic simulation methods have been applied to study small-length-scale hydrodynamics; these include Monte Carlo methods [5], the smooth particle applied mechanics (SPAM) method [111], lattice Boltzmann/lattice gas automata [42, 113], and the direct simulation Monte Carlo (DSMC) method [20], most of which are mainly applicable for gasses. Molecular dynamics is highly versatile and plays an ever increasing role in studying nanoscale phenomena [26], and we will apply only this simulation method. The following is not meant to be a thorough discussion of the molecular dynamics technique, but rather an introduction to the necessary terminology that will be used in the remaining chapters, and, importantly, also a justification for the use of molecular dynamics to explore nanoscale hydrodynamics. The interested reader is referred to the classical books on the subject, for example, Refs. [5, 74, 184].

Molecular dynamics is founded in classical mechanics, wherein Newton's second law is integrated numerically for each particle in the system. Here a particle can be a single atom, a molecule, or a group of atoms that move in a coherent fashion. If $\mathbf{F}_i$ is the force acting on particle i, having position $\mathbf{r}_i$, velocity $\mathbf{v}_i$, and mass m_i, the equations of motion are

$$\frac{d\mathbf{r}_i}{dt} = \mathbf{v}_i \text{ and } m_i \frac{d\mathbf{v}_i}{dt} = \mathbf{F}_i. \tag{1.21}$$

Then, for N particles we solve $6N$ coupled and, in general, nonlinear differential equations. In molecular dynamics we obviously assume that quantum effects can be safely ignored. As we have noticed, at ambient conditions this will indeed be the case for most liquids and fluids, as the de Broglie wavelength is less than one angstrom and significantly smaller than the relevant length scale. Not all phenomena can be described correctly by classical theories; for example, quantum mechanical effects occurring inside the wall may be relevant for the wall–fluid interactions and therefore also the hydrodynamics in highly confined geometries. These systems are not treated here.

Molecular dynamics relies on accurate models for the particle interactions; these inter-particle interactions are conservative and are therefore often given through a potential function U. In standard simulations we assume that the particles are spherical symmetric point masses and that the interactions are only pairwise such that $\mathbf{F}_i^c = -\nabla U(r_{ij})$, where $\mathbf{F}_i^c$ is used to underline that the force is conservative and r_{ij} is the distance between particle i and particle j. The famous Lennard–Jones pair potential reads

$$U_{LJ}(r_{ij}) = 4\varepsilon \left[\left(\frac{\sigma}{r_{ij}} \right)^{12} - \left(\frac{\sigma}{r_{ij}} \right)^{6} \right], \tag{1.22}$$

where ε and σ represent the interaction strength and characteristic diameter of the particles, respectively; most importantly here is that σ is typically on the order of a few angstroms. We are already now running out of symbols and do not want to confuse the length scale symbol with the standard deviation, or the energy scale with the dielectric constant. The first term is a repulsive term which accounts for the force due to electron repulsion at small inter-particle distances. The second term models the induced dipole moment; it is longer ranged and attractive. This is also known as the London dispersion force.

We will often use molecular dynamics simulations of methane fluid as controlled numerical experiments to test the hydrodynamic theories. The molecule is approximated to be a point mass spherical molecule – by far most of the mass is located in the carbon atom nucleus. The intermolecular interactions are modelled via the Lennard–Jones potential, where $m = 16$ g/mol, $\varepsilon/k_B = 148$ K, and $\sigma = 3.7$ Å [155]. The results from the molecular dynamics simulations can be presented in standard SI-units, but sometimes it is more convenient and insightful to list the results in units of σ, ε, and molecular mass m_i. Importantly, the unit of time in this unit system is $\sigma\sqrt{m/\varepsilon}$ and length is σ. We will use both SI units and molecular dynamics unit; the latter we denote MD units.

Table 1.1 Self-diffusivity, D_s, shear viscosity, η_0, relative permittivity, ε_r, heat of vaporisation, ΔH_{vap}, isothermal expansivity, χ_T, Debye relaxation time, τ_D, and thermal expansivity, α_p, for the SPC/Fw water model under ambient conditions. Both molecular dynamics results and the corresponding experimental values are listed. From Ref. [213].

	Units	SPC/Fw (MD)	Exp.
D_s	10^{-9} m^2 s^{-1}	2.35 ± 0.05	2.3
η_0	10^{-3} Pa s	0.75	0.85
ε_r		80 ± 2	78.5
ΔH_{vap}	kcal mol^{-1} K^{-1}	10.7 ± 0.1	10.52
χ_T	10^{-5} atm^{-1}	4.50	4.58
τ_D	psec.	9.5	8.3
α_p	10^{-4} K^{-1}	4.98	2.0

A lot of general information can be obtained by studying point mass particles like methane. However, some phenomena require that we use a more detailed model for the molecules, and today the molecular dynamics community simulates complex molecular systems with advanced interaction models; see Sadus [186] and Leach [142] for an overview. In general, we can write up a force field model that also includes Coulomb interactions, covalent bonds, forces due to angles and dihedral angles, and so forth. In terms of the potential function this is written as

$$U = U_{\text{LJ}} + U_{\text{coulomb}} + U_{\text{bonds}} + U_{\text{angles}} + U_{\text{dihedral}} + \ldots . \qquad (1.23)$$

Widely used models are, for example, the CHARMM [36] and OPLS [122] force fields that give explicit expressions for the different terms and model parameter values depending on the specific systems under investigation. Again, we will not go into detail with the different interaction models, as this is far outside the scope of the text. It is, however, important to highlight that molecular dynamics can, with an accurate interaction model, predict the different mechanical, dynamical, and thermodynamic properties quite well under normal pressures and temperatures, where nanoscale fluid systems often operate. As an example of this, Table 1.1 lists different physical coefficients calculated from molecular dynamics simulations at equilibrium for the flexible simple point charge water model (SPC/Fw) [181, 213]. For comparison purposes, the corresponding experimental values are also given. Except for the thermal expansivity, the model results agree well with the experimental measured values. This indicates that molecular dynamics indeed can capture many of the underlying physical processes correctly, including the processes relevant for hydrodynamics. Importantly, water is not easily modelled, as the different properties of the liquid are a result of the complex long-ranged hydrogen-bond network.

Since we are solving a very large set of coupled differential equations, the number of particles and the time we can reach are both limited. In molecular dynamics large

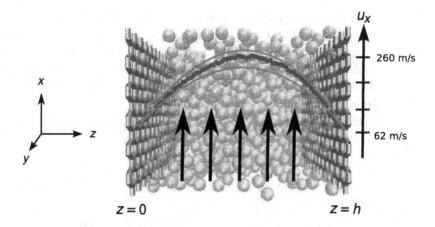

Figure 1.9   Molecular dynamics simulation of a planar Poiseuille flow. Symbols represent simulation data for the streaming velocity, and lines are the Navier–Stokes predictions; the two lines indicate the extremes in the uncertainty coming from the viscosity. Arrows illustrate the external force. Reprinted with permission from Ref. [102].

implies small: in 2013, the SuperMUC supercomputer simulated an impressively large system of 4.125×10^{12} particles, but here each time step (corresponding to approximately a femtosecond) took the computer 40 seconds, seriously limiting the time scale that can be studied. In the other extreme, one can reach 2×10^4 integration times steps per second for a small system size of 10^3 particles using Graphical Processor Units (GPUs) [9]. In the context of nanoscale hydrodynamics the number of particles is usually not too critical. For example, to simulate a methane fluid flow in a slit-pore geometry of height 10 nm we will need around 27×10^3 methane molecules if the simulation box is a perfect cube. For water in the same geometry we need approximately 10^5 hydrogen and oxygen atoms. The problem often lies in reaching realistic times, especially for charged systems like water, where the long-ranged Coulomb interactions are very computationally demanding. Even with a small number of particles, the time reached, τ_{obs}, with current computers usually does not exceed 10–100 nanoseconds. Thus, the phenomena we study with molecular dynamics must have small characteristic time scales. Often we must apply large external forces in order to excite the relevant physical mechanisms needed to reach a sufficiently small Deborah number, $\text{De} = \tau_s / \tau_{\text{obs}}$.

Figure 1.9 shows a snapshot from a molecular dynamics simulation of a methane fluid flowing in a slit-pore geometry similar to the one in Fig. 1.1. The molecular interaction is modelled through the Lennard–Jones potential, Eq. (1.22). The confining walls are also composed of Lennard–Jones particles and are positioned in a graphene-type lattice. The flow is generated by applying a constant external force that acts on the centre of mass of the methane molecules, and the resulting viscous heating is removed by applying a thermostat. In this manner a planar Poiseuille flow is simulated.

Even if equilibrium molecular dynamics results for the transport properties agree well with the experimental data, it is by no means trivial that the hydrodynamic model, in this case the Navier–Stokes equation, and the non-equilibrium simulations agree.

To make a direct comparison for the system in Fig. 1.9, we note that $h = 3.3$ nm, the density is $\rho = 270$ kgm^{-3}, the shear viscosity is $\eta_0 = 9.3 \pm 0.6$ µPa·s, and that the external acceleration applied is a staggering $g = 5.0 \times 10^{12}$ ms^{-1}. Using $U \approx 100$ ms^{-1} as the characteristic velocity, the Reynolds number, Eq. (1.2), is around 10. Hence, the flow is laminar. Furthermore, as the speed of sound for methane at this state point is approximately 10^3 ms^{-1}, the Mach number $M = U/c_s < 0.3$ and we need not consider fluid compressibility effects. In the slit-pore geometry the Navier–Stokes equation is reduced to a tractable boundary value problem

$$\eta_0 \frac{d^2 u_x}{dz^2} + \rho g = 0, \qquad (1.24)$$

with

$$u_x(0) = u_x(h) = u_w = 62 \text{ ms}^{-1}, \qquad (1.25)$$

where u_w is the slip velocity at the wall. The solution for this problem gives the well-known Poiseuille flow quadratic profile

$$u_x(z) = \frac{\rho g}{2\eta_0} z(h - z) + u_w. \qquad (1.26)$$

The prediction from the Navier–Stokes equation is also shown in the figure (lines), and is in very good agreement with the time-averaged simulation data (filled circles). This is a crucial result; the hydrodynamic prediction, Eq. (1.26), where molecular details are strictly not considered agrees with the (time-averaged) molecular simulation results coming from solving the Newtonian equation of motion for each molecule, Eq. (1.21). The scenario that both of these very different descriptions of the flow are incorrect yet still produce the same result is highly unlikely. While definitely not being a proof, this example illustrates that hydrodynamics can for simple systems be applied on the nanoscale, *and* molecular dynamics can be applied to perform idealised numerical experiments of nanoscale fluid systems. Now, almost all real nanoscale fluid systems are not as simple and idealised as this, and laboratory experiments must, of course, always be the final test of our theoretical predictions.

As we have mentioned, the time scales we can reach with molecular dynamics are small compared to typical hydrodynamic time scales. In our Poiseuille flow simulation the external acceleration applied to drive the flow is of literally astronomical magnitude. This is necessary in order to obtain a well-developed velocity profile within the nanoscale time frame available. This large acceleration produces unrealistically large streaming velocities and strain rates, and yet, the simulation data agree with the Navier–Stokes predictions. The reason for this lies in the fact that the local Newtonian viscosity law for shear stress, Eq. (1.15), applies, that is, the system response is still linear and local. Lennard–Jones-type systems show a strain-rate-independent viscosity (Newtonian behaviour) for strain rates less than 10^{10}–10^{11} s^{-1} in the liquid phase [200] which is the same order of magnitude as the flow in Fig. 1.9. Non-Newtonian effects must always be considered, and for more complex fluids the regime where Newton's viscosity law is valid may not be accessible by molecular dynamics even when using highly optimised algorithms and hardware [146].

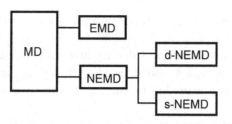

Figure 1.10   Schematic diagram of the three standard molecular dynamics (MD) techniques. Equilibrium (EMD), direct non-equilibrium (d-NEMD), and synthetic non-equilibrium (s-NEMD) simulations.

1.3.1 Molecular Dynamics Simulation Techniques

Simulations in molecular dynamics are divided into two main categories, equilibrium simulations (EMD) and non-equilibrium simulations (NEMD). With EMD we simulate the system in some well-known statistical mechanical ensemble, for example, the microcanonical ensemble where the number of molecules, volume, and energy are constants along the phase space trajectory. In this way we can use classical statistical mechanical results for the particular ensemble to derive the given properties we are studying. The properties listed in Table 1.1 are an example of a simulation in the canonical ensemble, where the number of molecules, volume, and temperature are constants. In Chapters 3 and 4 we rely heavily on equilibrium simulations to test how hydrodynamics predict relaxation phenomena in equilibrium.

In Chapters 5 and 6 we explore non-equilibrium systems. In non-equilibrium we can perform either direct (d-NEMD) or synthetic (s-NEMD) simulations. For d-NEMD we try to mimic the real physical experiment, at least to some approximation. Our Poiseuille flow is an example of this. The fluid flow is generated by application of some external driving force, $\mathbf{F}_i^{\text{ext}} = m_i \mathbf{g}$, by moving a wall or similar, and the resulting viscous heating is removed by thermostating the confining wall atoms; a discussion of the different thermostating methods can be found in Ref. [18]. The d-NEMD method is relatively straightforward to implement; for example, the driving force can be added directly to Eq. (1.21), so that for the fluid particles we have

$$\frac{\mathrm{d}\mathbf{r}_i}{\mathrm{d}t} = \mathbf{v}_i \text{ and } m_i \frac{\mathrm{d}\mathbf{v}_i}{\mathrm{d}t} = \mathbf{F}_i = \mathbf{F}_i^{\text{ext}} + \mathbf{F}_i^c. \tag{1.27}$$

However, d-NEMD is not always suitable if one needs to study and isolate a specific fluid phenomenon, as the confining walls often complicate and clutter the problem. Another problem with d-NEMD is that the statistical mechanics for such systems is not well developed, and we need to approach the analysis of our simulation results with care.

To overcome the problem associated with d-NEMD, one can perform s-NEMD simulations. Here the equations of motion are changed in order to probe a specific dynamical feature. The particles do not follow the simple Newtonian equations of motion, but their dynamics are changed synthetically such that the system possesses, for example, a constant [65, 145] or spatially varying [12, 80] strain rate while keeping the local density and temperature constant (on average). s-NEMD is for this reason also referred to as homogeneous NEMD. In Section 1.1.5 this technique was used to investigate the fluid response to an imposed sinusoidal strain rate.

It is important to note that while the equations of motion are synthetic (also referred to as fictitious) and often not realisable in the laboratory, the system trajectory follows Gauss' principle of least constraints [66], and we can expect that the s-NEMD method probes the correct physics. The hydrodynamic equations derived from s-NEMD equations of motion are effected by the thermostat and are not the same as those derived directly from Eq. (1.21). Therefore, great care must be taken when interpreting and analysing results from such simulations [200]. Again, it is outside the scope of this text to pursue an in-depth introduction to the different MD techniques. For more details on how to implement the d-NEMD and s-NEMD methods, the reader is referred to the books by Evans and Morriss [66] and Todd and Daivis [200].

1.3.2 Mesoscale Molecular Dynamics

In order to extend the time scales we can reach with standard molecular dynamics methods, that is, increase τ_{obs} and reduce the Deborah number, alternative simulation methods have been devised. One such method is dissipative particle dynamics (DPD) [63, 112]. Rather than solving Newton's equations of motion for the individual atom or molecule, DPD solves the equation of motion for a collection of particles moving in a coherent fashion. This coherent motion is described through a single DPD particle, and in order to account for the coarse graining, the random force, $\mathbf{F}_i^R$, and dissipative force, $\mathbf{F}_i^D$ terms are augmented to the Newtonian equation of motion, that is,

$$\frac{d\mathbf{r}_i}{dt} = \mathbf{v}_i \text{ and } m_i \frac{d\mathbf{v}_i}{dt} = \mathbf{F}_i^c + \mathbf{F}_i^R + \mathbf{F}_i^D, \tag{1.28}$$

in the absence of any external forces. Importantly, these two forces are defined such that the total momentum is conserved, and this makes DPD fundamentally different from Brownian simulations and ensures hydrodynamic conservation of momentum. As with standard molecular dynamics, the force $\mathbf{F}_i^c$ represents the conservative interactions with other particles; but due to the coarse graining, these interactions are usually modelled as being 'soft', allowing the DPD particles to overlap; and a larger integration time step can be applied, which is important if we wish to simulate the system for a larger period of time. DPD is then based on a set of stochastic differential equations and is a mesoscopic description of the fluid. It is not always straightforward to extract the physical time and length scales in such simulations, that is to say, what size a DPD particle has. Moreover, while it has been shown that the balance equations for mass and momentum densities are obeyed, the total energy density is not conserved due to the random and dissipative forces [154]. However, it has been shown that DPD do capture many of the underlying hydrodynamics processes [103]; it is therefore a very potent alternative to classical molecular dynamics simulations, and we will use DPD to explore viscoelastic phenomena.

2 Balance Equations

This chapter introduces a microscopic formalism to derive the balance equation for any hydrodynamic variable. This can be, as we shall see, the balance equation for mass density, momentum density, or kinetic temperature. The formalism is based on the molecular definitions of the variables, and is therefore fundamentally different from the classical macroscopic treatment used in standard textbooks [15, 38, 207]. One advantage of the formalism is that the terms entering the equations are associated with molecular quantities and thus provide the link between the detailed microscopic and coarse-grained continuum pictures. Derivation of the balance equations based on the molecular quantities is not new; see, for example, more recent treatments in [66, 90, 102, 195, 200]. Here it is introduced in a more general setting and illustrated with several examples.

Even if the terms in the balance equations are expressed through the molecular quantities, the balance equations do not form closed mathematical problems, as the terms are still unknown functions of the hydrodynamic variables. One therefore introduces constitutive relations between the system fluxes and the corresponding system forces. The forces are typically given by gradients of the variables themselves; Newton's law of viscosity, Eq. (1.15), is an example of such a relation. The constitutive relations are models. Usually, the relations are local in time and space in the sense that the flux is dependent only on the force at that particular point and at that instant in time. Spatio-temporal correlation effects are therefore ignored. Moreover, one typically assumes linearity, that is, the fluxes are linearly dependent on the forces. Applying the constitutive relations leads to closed mathematical problems, which are solvable through either mathematical analysis or standard numerical methods. We shall see examples of this throughout the text, but in this chapter we start by deriving the underlying balance equations.

Let us sidetrack a bit already and introduce how we characterise the hydrodynamic variables; this leads to the concept of tensors. If you are a mathematician, you may associate a tensor with a mapping from one abstract mathematical structure to another, say, from a vector space into the real numbers (i.e., a scalar product). The formal theory on tensors is rather involved, but we need not go through the details here. For our purpose it suffices to use a much more informal definition: a tensor is a physical property or quantity which is independent of the choice of coordinate system. For example, the velocity at a given point does not change because we change the coordinate system; only how we represent the velocity components depends on the specific coordinate system. The moment of inertia is another example: in the fixed (or lab) frame of reference

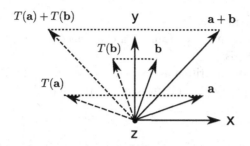

Figure 2.1
The reflection of vectors **a**, **b**, and **a** + **b**.

it is represented by a symmetric 3×3 array, and in the so-called principal coordinate system it can be reduced to a diagonal array.

The tensor rank is defined by how the tensor is represented, such that a rank-0 tensor is represented by a scalar, a rank-1 tensor by a one-dimensional number array (a vector), rank-2 by a two-dimensional array (analogous to how we represent a matrix), and so forth. Here we use the terms scalar and vector for rank-0 and rank-1 tensors, respectively, and otherwise we use the more general tensor term specifying the rank.

In our treatment here we only distinguish between quantities that are tensors and pseudo-tensors. Pseudo-tensors can be formed from a binary operation of two tensors. For example, the cross-product of two vectors results in a pseudo-vector; but addition of two vectors, on the other hand, results in a vector. One way to illustrate the distinction between a tensor and a pseudo-tensor is by how they behave under reflection. Let **a** and **b** be vectors and T be a linear map that reflects a vector about the y-axis; T can be represented as a simple matrix that flips the sign of the x-vector components. We first let the vector $\mathbf{c} = \mathbf{a} + \mathbf{b}$ and we then have from the linearity of T that $T(\mathbf{c}) = T(\mathbf{a} + \mathbf{b}) = T(\mathbf{a}) + T(\mathbf{b})$. Thus, the reflection of the vector **c** is given by the sum of the reflection of **a** and **b**. See Fig. 2.1. Now, if $\mathbf{c} = \mathbf{a} \times \mathbf{b}$, which is parallel to the z-axis, we have from the right-hand rule $T(\mathbf{c}) = T(\mathbf{a} \times \mathbf{b}) = -T(\mathbf{a}) \times T(\mathbf{b})$; that is, the reflection of **c** is given by the cross-product of the reflections of **a** and **b**, but with an additional sign flip and **c** is a pseudo-vector. To make the distinction clear, tensors are often referred to as polar-tensors, but we will simply write tensor.

We will deal mostly with tensors and pseudo-tensors in three-dimensional Cartesian coordinates. The components for vectors and pseudo-vectors are identified through the subscripts x, y, and z. The components of rank-2 tensors (and pseudo-tensors) will be given subscripts xx, xy, xz, and so on.[1] A quantity of rank-0 or of unknown rank is given by an ordinary mathematical symbol, and rank-1 and above are denoted with boldface.

Let us return to the main track again. If $\mathbf{r} = (x, y, z)$ denotes the position vector and t the time, then we write a hydrodynamic quantity as $A = A(\mathbf{r}, t)$. A can be a scalar quantity, say, the mass density; a vector quantity, say, the momentum density; or even a pseudo-vector like the vorticity. A is assumed to be 'well' behaved, that is, the derivatives

---

[1] So we do not distinguish between covariant and contravariant tensors.

with respect to time and space exist, and it has a Fourier transform. A can then be described mathematically as a field variable; this is in accordance with the continuum hypothesis.

A is often written as a product of the mass density, $\rho = \rho(\mathbf{r},t)$, and the associated field variable per unit mass, $\phi = \phi(\mathbf{r},t)$, namely,

$$A(\mathbf{r},t) = \rho(\mathbf{r},t)\phi(\mathbf{r},t). \tag{2.1}$$

From here on the explicit position and time dependencies are given only if they provide important information or for clarity. We will explore systems where the rate of change of A is due to three different processes: (i) a production process (for example, an external force field can be applied generating linear momentum), (ii) an advective process wherein A is carried along the bulk fluid streaming motion, and (iii) a diffusive process which tends to remove any gradients, that is, to homogenise the system. The general expression for the rate of change of A is written in the differential form [52],

$$\frac{\partial \rho \phi}{\partial t} = \sigma - \nabla \cdot (\rho \mathbf{u} \phi) - \nabla \cdot \mathbf{J}, \tag{2.2}$$

where σ is the production term, $\mathbf{u}$ is the streaming velocity, and $\mathbf{J}$ is the flux of A, that is, the rate at which A flows through a unit surface area. ∇ is the usual del (or nabla) operator known from vector calculus. In Cartesian coordinates this is

$$\nabla = \left(\frac{\partial}{\partial x}, \frac{\partial}{\partial y}, \frac{\partial}{\partial z} \right), \tag{2.3}$$

and $\nabla \cdot \mathbf{J}$ is then the divergence of $\mathbf{J}$. Equation (2.2) is the general balance equation for A, accounting for the three processes described earlier. A special case is for the mass density, where we have $\phi = 1$ and $\sigma = 0$.

Importantly, each term in the balance equation must have same tensorial character. This means that they must have the same tensorial rank and must transform the same way; hence, we cannot add, for example, vectors and pseudo-vectors. Also, having introduced the del operator, we can list the following important results concerning tensors that we need for later:

$\nabla \times \mathbf{a}$ is a pseudo-vector if $\mathbf{a}$ is a vector,

$\nabla \times \mathbf{a}$ is a vector if $\mathbf{a}$ is a pseudo-vector, and

$\mathbf{a} \times \mathbf{b}$ is a vector if $\mathbf{a}$ is a pseudo-vector and $\mathbf{b}$ is a vector. $\qquad$ (2.4)

Instead of studying the balance equation directly in real space, it can be advantageous to study the dynamics in Fourier space, or specifically, the dynamics of the Fourier coefficients. To this end, recall that the Fourier coefficients can be found from the (three-dimensional) Fourier transform, $\mathcal{F}$, acting on a function f as

$$\mathcal{F}[f(\mathbf{r},t)] = \tilde{f}(\mathbf{k},t) = \int_{-\infty}^{\infty} f(\mathbf{r},t) e^{-i\mathbf{k}\cdot\mathbf{r}} \, d\mathbf{r}, \tag{2.5}$$

where $\mathbf{k} = (k_x, k_y, k_z)$ is the wavevector. Notice that we are using an abbreviation format here; the integral in Eq. (2.5) represents the volume integral over all three Cartesian coordinates and therefore, $d\mathbf{r} = dxdydz$. See the Appendix for further clarifications. From integration by parts, two useful properties of the Fourier transform can be shown:

$$\mathcal{F}[\nabla \cdot f(\mathbf{r},t)] = i\mathbf{k} \cdot \widetilde{f}(\mathbf{k},t) \tag{2.6a}$$

$$\mathcal{F}[\nabla \times f(\mathbf{r},t)] = i\mathbf{k} \times \widetilde{f}(\mathbf{k},t). \tag{2.6b}$$

From Fourier transforming Eq. (2.2), one obtains the equation for the Fourier coefficients; using the identity in Eq. (2.6a),

$$\frac{\partial}{\partial t}\widetilde{\rho\phi}(\mathbf{k},t) = \widetilde{\sigma}(\mathbf{k},t) - i\mathbf{k} \cdot \widetilde{\rho\mathbf{u}\phi}(\mathbf{k},t) - i\mathbf{k} \cdot \widetilde{\mathbf{J}}(\mathbf{k},t), \tag{2.7}$$

due to the linear properties of $\mathcal{F}$. We refer to this as the balance equation in Fourier space.

The balance equation can be studied in the limit of zero wavevector, $\mathbf{k} \rightarrow \mathbf{0}$, or equivalently on large wavelengths. To this end, all the terms on the right-hand side are expanded and written in the following form:

$$\widetilde{\sigma}(\mathbf{k},t) = \widetilde{\sigma_0}(t) + \widetilde{\sigma_1}(\mathbf{k},t) + \dots \tag{2.8a}$$

$$\widetilde{\rho\mathbf{u}\phi}(\mathbf{k},t) = (\widetilde{\rho\mathbf{u}\phi})_0(t) + (\widetilde{\rho\mathbf{u}\phi})_1(\mathbf{k},t) + \dots \tag{2.8b}$$

$$\widetilde{\mathbf{J}}(\mathbf{k},t) = \widetilde{\mathbf{J}}_0(t) + \widetilde{\mathbf{J}}_1(\mathbf{k},t) + \dots, \tag{2.8c}$$

where $\widetilde{\sigma}_0$ is the zeroth-order term in the expansion for the production term, $\widetilde{\sigma}_1$ is the first-order term, and so forth. To first order the balance equation reads

$$\frac{\partial}{\partial t}\widetilde{\rho\phi}(\mathbf{k},t) = \widetilde{\sigma_0}(t) + \widetilde{\sigma_1}(\mathbf{k},t) - i\mathbf{k} \cdot (\widetilde{\rho\mathbf{u}\phi})_0(t) - i\mathbf{k} \cdot \widetilde{\mathbf{J}}_0(t), \tag{2.9}$$

and for zero wavevector the balance equation is

$$\frac{\partial}{\partial t}\widetilde{\rho\phi}(\mathbf{k},t) = \widetilde{\sigma_0}(t), \tag{2.10}$$

as $\sigma_1 = 0$ for $\mathbf{k} = \mathbf{0}$. The hydrodynamic quantity A is said to be a globally conserved quantity if the wavevector-independent part of the production term is zero, that is, $\widetilde{\sigma_0} = 0$ for all t.

The programme is now to derive the balance equation for the Fourier coefficients, Eq. (2.7), from fundamental molecular definitions of A. Once this is done, we can infer the real space balance equation in the form of Eq. (2.2), having molecular or microscopic interpretations of the relevant fluxes.

2.1 The Microscopic Operators

Intuitively, A is the volume average of the corresponding molecular variable a in some small fluid element $\mathcal{V}$; see Eqs. (1.19) and (1.20). For simplicity assume that the fluid element is a sphere with centre $\mathbf{r}$ and radius ε. Let $\mathbf{r}_i$ be the molecular centre of mass. Then the molecule is located in $\mathcal{V}$ if the norm $||\mathbf{r} - \mathbf{r}_i|| \leq \varepsilon$; that is, we can define the step function

$$\Pi(\mathbf{r} - \mathbf{r}_i(t)) = \begin{cases} 1 & \text{if } ||\mathbf{r} - \mathbf{r}_i|| \leq \varepsilon \\ 0 & \text{otherwise.} \end{cases} \tag{2.11}$$

In general, one can chose any fluid element geometry and from that define the appropriate norm. The hydrodynamic variable A is then defined from the corresponding molecular quantity, a, by

$$A(\mathbf{r}, t) = \frac{1}{\Delta\mathcal{V}} \sum_i a_i(t)\Pi(\mathbf{r} - \mathbf{r}_i(t)), \tag{2.12}$$

where $\Delta\mathcal{V}$ is the volume of $\mathcal{V}$. In the limit of $\Delta\mathcal{V} \to 0$ we write this in terms of the Dirac delta:

$$A(\mathbf{r}, t) = \sum_i a_i(t)\,\delta(\mathbf{r} - \mathbf{r}_i(t)). \tag{2.13}$$

Strictly, from Eq. (2.13) the hydrodynamic variables will be zero at points where there are no particles and diverge at points $\mathbf{r} = \mathbf{r}_i$. Hence, over time, A at some point will fluctuate between zero and infinite, which is not really meaningful and definitely not in accordance with the continuum hypothesis. However, this definition of the distribution is extremely powerful and unambiguous when the Dirac delta is used together with an integral over a fluid volume, as we will see soon. The Dirac delta has units of inverse its argument, here inverse length cubed (inverse volume). Also, see the Appendix for the three-dimensional Dirac delta and its properties. Evans and Morriss [66] refer to Eq. (2.13) as the instantaneous microscopic definition of A, and we will use this throughout the book.

The advantage of writing the limit of Eq. (2.12) in terms of the Dirac delta is evident when Fourier transforming Eq. (2.13),

$$\widetilde{A}(\mathbf{k}, t) = \widetilde{\rho\phi}(\mathbf{k}, t) = \sum_i a_i(t) \int_{-\infty}^{\infty} \delta(\mathbf{r} - \mathbf{r}_i)\, e^{-i\mathbf{k}\cdot\mathbf{r}}\, d\mathbf{r}$$
$$= \sum_i a_i(t)\, e^{-i\mathbf{k}\cdot\mathbf{r}_i}, \tag{2.14}$$

using the properties of the Dirac delta; again see the Appendix. The derivative of Eq. (2.14) with respect to time follows:

$$\frac{\partial}{\partial t}\widetilde{\rho\phi}(\mathbf{k}, t) = \sum_i \frac{da_i}{dt} e^{-i\mathbf{k}\cdot\mathbf{r}_i} + a_i(-i\mathbf{k}\cdot\mathbf{v}_i)\, e^{-i\mathbf{k}\cdot\mathbf{r}_i}$$
$$= \sum_i \left(\frac{da_i}{dt} - i\mathbf{k}\cdot(\mathbf{v}_i a_i) \right) e^{-i\mathbf{k}\cdot\mathbf{r}_i}, \tag{2.15}$$

where $\mathbf{v}_i$ is the centre-of-mass velocity of molecule i. In Eq. (2.15) the identity

$$a(\mathbf{b}\cdot\mathbf{c}) = \mathbf{b}\cdot(\mathbf{c}a) \tag{2.16}$$

is used. This identity is, perhaps, trivial when a is a scalar quantity. When a is a vector (or pseudo-vector), $a = \mathbf{a}$, the product $\mathbf{ca}$ is an outer product (or dyadic) and the result is a rank-2 tensor with components

$$(ca)_{\alpha\beta} = c_\alpha a_\beta, \qquad (2.17)$$

where α and β runs over the x, y, z vector components.

From Eq. (2.17) we see that the outer product is not commutative in general, $\mathbf{ca} \neq \mathbf{ac}$. In the special case where the vectors $\mathbf{a}$ and $\mathbf{c}$ are parallel, commutation of the outer product is fulfilled, and the resulting rank-2 tensor is a symmetric tensor. To see this we write $\mathbf{a} = A\mathbf{c}$, where A is a real number, and we have $(\mathbf{ac})_{\alpha\beta} = Ac_\alpha c_\beta = Ac_\beta c_\alpha$ or, in vector notation, $\mathbf{ac} = (\mathbf{ac})^T = \mathbf{ca}$, where the superscript T means the tensor transpose.

The molecular velocity can be decomposed into a peculiar or thermal part, $\mathbf{c}$, and an advective part, $\mathbf{u}$, [66, 195]:

$$\mathbf{v}_i = \mathbf{c}_i + \mathbf{u}(\mathbf{r}_i, t). \qquad (2.18)$$

The advective part is the mass-weighted average fluid velocity (streaming velocity) of the fluid element with centre of mass located at $\mathbf{r}_i$. The thermal and advective velocities are uncorrelated and from conservation of momentum we have $\sum_i m_i \mathbf{c}_i = \mathbf{0}$, where m_i is the mass of the molecule. The outer product is distributive, so we have $\mathbf{v}_i a_i = \mathbf{c}_i a_i + \mathbf{u}(\mathbf{r}_i, t) a_i$. Equation (2.15) is therefore written as

$$\frac{\partial}{\partial t} \widetilde{\rho\phi}(\mathbf{k}, t) = \sum_i \left(\frac{da_i}{dt} - i\mathbf{k} \cdot \mathbf{c}_i a_i - i\mathbf{k} \cdot \mathbf{u}(\mathbf{r}_i, t) a_i \right) e^{-i\mathbf{k}\cdot\mathbf{r}_i}. \qquad (2.19)$$

If we compare this result with Eq. (2.7), we see that it is the equation for the Fourier coefficients of A, but expressed in terms of molecular quantities.

The right-hand side of Eq. (2.19) defines the operator,

$$\mathcal{G}[a_i] = \sum_i \left(\frac{da_i}{dt} - i\mathbf{k} \cdot \mathbf{c}_i a_i - i\mathbf{k} \cdot \mathbf{u}(\mathbf{r}_i, t) a_i \right) e^{-i\mathbf{k}\cdot\mathbf{r}_i}, \qquad (2.20)$$

and we can write Eq. (2.19) in terms of this operator, namely,

$$\frac{\partial}{\partial t} \widetilde{\rho\phi}(\mathbf{k}, t) = \mathcal{G}[a_i]. \qquad (2.21)$$

From the $\mathcal{G}$-operator we can also express the dynamics of A in the small wavevector limit. First, expanding the exponential function around zero wavevector gives

$$e^{-i\mathbf{k}\cdot\mathbf{r}_i} = 1 - i\mathbf{k} \cdot \mathbf{r}_i - \frac{1}{2}(\mathbf{k} \cdot \mathbf{r}_i)^2 + \dots. \qquad (2.22)$$

Substituting Eq. (2.22) into Eq. (2.20) and collecting the terms with respect to wavevector, one has

$$\mathcal{G}[a_i] = \sum_i \frac{da_i}{dt} - i\mathbf{k} \cdot \sum_i \mathbf{r}_i \frac{da_i}{dt} - i\mathbf{k} \cdot \sum_i \mathbf{c}_i a_i - i\mathbf{k} \cdot \sum_i \mathbf{u}(\mathbf{r}_i, t) a_i$$
$$- \frac{1}{2} \sum_i (\mathbf{k} \cdot \mathbf{r}_i)^2 \frac{da_i}{dt} - \sum_i (\mathbf{k} \cdot \mathbf{r}_i)\mathbf{k} \cdot (\mathbf{c}_i a_i + \mathbf{u}(\mathbf{r}_i, t) a_i) + \dots. \qquad (2.23)$$

From this it is convenient to define an operator up to first order in the wavevector

$$\mathcal{H}[a_i] = \sum_i (1 - i\mathbf{k} \cdot \mathbf{r}_i) \frac{da_i}{dt} - i\mathbf{k} \cdot \sum_i \mathbf{c}_i a_i - i\mathbf{k} \cdot \sum_i \mathbf{u}(\mathbf{r}_i, t) a_i, \qquad (2.24)$$

such that the dynamics in the small wavevector limit is given by

$$\frac{\partial}{\partial t}\widetilde{\rho\phi}(\mathbf{k},t) = \mathcal{H}[a_i]. \quad \text{(small } \mathbf{k})$$ (2.25)

The $\mathcal{H}$-operator has been denoted the microscopic hydrodynamic operator [102], and we will adopt this name. As we will see, the balance equations can often be derived by studying only the small wavevector limit and simply applying Eq. (2.24).

2.2 Application of the Operators

Hydrodynamics considers the balance equations for mass, momentum, and energy densities. In this section we will use the operators $\mathcal{G}$ and $\mathcal{H}$ to derive these balance equations; however, the classical total energy balance is replaced with the balance equation for the thermal kinetic energy, as this readily gives the equation for the kinetic temperature that we use in the subsequent chapters.

2.2.1 Mass Balance

The mass density is defined from the molecular masses m_i [90]:

$$\rho(\mathbf{r},t) = \sum_i m_i \delta(\mathbf{r} - \mathbf{r}_i).$$ (2.26)

Comparing with Eq. (2.13), we identify $a_i = m_i$. The dynamics given by the $\mathcal{G}$-operator is

$$\begin{aligned}
\mathcal{G}[m_i] &= \sum_i \left(\frac{\mathrm{d}m_i}{\mathrm{d}t} - i\mathbf{k}\cdot m_i\mathbf{c}_i - i\mathbf{k}\cdot m_i\mathbf{u}(\mathbf{r}_i,t) \right) e^{-i\mathbf{k}\cdot\mathbf{r}_i} \\
&= -i\mathbf{k}\cdot\sum_i m_i\left(\mathbf{c}_i + \mathbf{u}(\mathbf{r}_i,t)\right) e^{-i\mathbf{k}\cdot\mathbf{r}_i} \\
&= -i\mathbf{k}\cdot\sum_i m_i\mathbf{v}_i e^{-i\mathbf{k}\cdot\mathbf{r}_i},
\end{aligned}$$ (2.27)

as the mass of each molecule is constant. The equation for the Fourier coefficients is therefore

$$\frac{\partial\widetilde{\rho}}{\partial t} = -i\mathbf{k}\cdot\sum_i m_i\mathbf{v}_i e^{-i\mathbf{k}\cdot\mathbf{r}_i}.$$ (2.28)

Equation (2.28) provides the microscopic definition of the linear momentum density (or mass current), $\mathbf{j}$. To see this, let

$$\mathbf{j}(\mathbf{r},t) = \rho(\mathbf{r},t)\mathbf{u}(\mathbf{r},t) = \sum_i m_i\mathbf{v}_i(t)\delta(\mathbf{r} - \mathbf{r}_i).$$ (2.29)

Taking the divergence and then Fourier transforming, we obtain

$$i\mathbf{k}\cdot\widetilde{\rho\mathbf{u}} = i\mathbf{k}\cdot\sum_i m_i\mathbf{v}_i e^{-i\mathbf{k}\cdot\mathbf{r}_i},$$ (2.30)

and we can write Eq. (2.28) as

$$\frac{\partial \widetilde{\rho}}{\partial t} = -i\mathbf{k} \cdot \widetilde{\rho \mathbf{u}}.$$ (2.31)

This is the Fourier transform of the well-known mass balance (or continuity) equation:

$$\frac{\partial \rho}{\partial t} = -\boldsymbol{\nabla} \cdot (\rho \mathbf{u}) = -\boldsymbol{\nabla} \cdot \mathbf{j}.$$ (2.32)

Hence we have shown that the definition of $\mathbf{j}$, Eq. (2.29), is consistent with the continuity equation.

Note that for zero wavevector the rate of change of the mass density is zero, meaning that mass is a conserved quantity.

From the $\mathcal{G}$-operator, the linear momentum is given by the total molecular velocity, that is, the sum of the thermal and streaming parts. An important note here: if we apply the $\mathcal{H}$-operator, we get

$$\mathcal{H}[m_i] = \sum_i (1 - i\mathbf{k} \cdot \mathbf{r}_i)\frac{\mathrm{d}m_i}{\mathrm{d}t} - i\mathbf{k} \cdot \sum_i m_i \mathbf{c}_i - i\mathbf{k} \cdot \sum_i m_i \mathbf{u}(\mathbf{r}_i, t)$$
$$= -i\mathbf{k} \cdot \sum_i m_i \mathbf{u}(\mathbf{r}_i, t),$$ (2.33)

as, by definition, $\sum_i m_i \mathbf{c}_i = \mathbf{0}$. In terms of the molecular quantities and in the small wavevector limit, the dynamics are then given by

$$\frac{\partial \widetilde{\rho}}{\partial t} = -i\mathbf{k} \cdot \sum_i m_i \mathbf{u}(\mathbf{r}_i, t).$$ (2.34)

This result is different from Eq. (2.28), and the mass balance equation must be derived from the $\mathcal{G}$-operator, as this correctly defines the linear momentum from the molecular velocity $\mathbf{v}$.

2.2.2 Linear Momentum Balance

We have now defined the linear momentum density $\mathbf{j}$ microscopically in Eq. (2.29); but before we apply the microscopic operators and derive the balance equation for this quantity, we need to show an important lemma regarding the uncorrelated velocities $\mathbf{c}_i$ and $\mathbf{u}$ [195].

The outer product of the streaming velocity follows the straightforward identities

$$\rho(\mathbf{uu}) = (\rho\mathbf{u})\mathbf{u} = \mathbf{u}(\rho\mathbf{u}).$$ (2.35)

Equivalently, the identities can be written in terms of the Fourier transforms

$$\mathcal{F}[\rho(\mathbf{uu})] = \mathcal{F}[(\rho\mathbf{u})\mathbf{u}] = \mathcal{F}[\mathbf{u}(\rho\mathbf{u})].$$ (2.36)

Having established the microscopic definitions of the mass and momentum densities, we can express each outer product in terms of the molecular quantities. First,

$$\mathcal{F}[\rho(\mathbf{uu})] = \mathcal{F}\left[\left(\sum_i m_i \,\delta(\mathbf{r} - \mathbf{r}_i)\right)\mathbf{u}(\mathbf{r}, t)\mathbf{u}(\mathbf{r}, t)\right]$$
$$= \sum_i m_i \mathbf{u}(\mathbf{r}_i, t)\mathbf{u}(\mathbf{r}_i, t)e^{-i\mathbf{k}\cdot\mathbf{r}_i}.$$ (2.37)

Then,

$$\mathcal{F}[(\rho\mathbf{u})\mathbf{u}] = \mathcal{F}\left[\left(\sum_i m_i \mathbf{v}_i\, \delta(\mathbf{r}-\mathbf{r}_i)\right)\mathbf{u}(\mathbf{r},t)\right]$$
$$= \sum_i m_i(\mathbf{c}_i\mathbf{u}(\mathbf{r}_i,t) + \mathbf{u}(\mathbf{r}_i,t)\mathbf{u}(\mathbf{r}_i,t))e^{-i\mathbf{k}\cdot\mathbf{r}_i}, \tag{2.38}$$

and finally,

$$\mathcal{F}[\mathbf{u}(\rho\mathbf{u})] = \mathcal{F}\left[\mathbf{u}(\mathbf{r},t)\sum_i m_i \mathbf{v}_i\, \delta(\mathbf{r}-\mathbf{r}_i)\right]$$
$$= \sum_i m_i(\mathbf{u}(\mathbf{r}_i,t)\mathbf{c}_i + \mathbf{u}(\mathbf{r}_i,t)\mathbf{u}(\mathbf{r}_i,t))e^{-i\mathbf{k}\cdot\mathbf{r}_i}. \tag{2.39}$$

Note, we do not have a microscopic definition of the streaming velocity $\mathbf{u}$, albeit later we will present an approximation for this. By equating Eqs. (2.37), (2.38), and (2.39), one has

$$\sum_i m_i \mathbf{c}_i \mathbf{u}(\mathbf{r}_i,t) = \sum_i m_i \mathbf{u}(\mathbf{r}_i,t)\mathbf{c}_i = \mathbf{0}, \tag{2.40}$$

that is, the sum of the outer product of the thermal and the streaming velocities is zero. Specifically, each component of the tensor is zero, and since the mass is non-zero we have

$$\sum_i c_{i,\alpha} u_\beta(\mathbf{r}_i,t) = \sum_i u_\alpha(\mathbf{r}_i,t)c_{i,\beta} = 0. \tag{2.41}$$

Contrary to the mass balance equation, we will derive the balance equation for linear momentum density by simply applying the $\mathcal{H}$-operator and show that this gives the correct form.

From Eq. (2.29) we readily identify $a_i = m_i\mathbf{v}_i$, and we have

$$\mathcal{H}[m_i\mathbf{v}_i] = \sum_i(1 - i\mathbf{k}\cdot\mathbf{r}_i)m_i\frac{d\mathbf{v}_i}{dt} - i\mathbf{k}\cdot\sum_i m_i\mathbf{c}_i\mathbf{v}_i - i\mathbf{k}\cdot\sum_i m_i\mathbf{u}(\mathbf{r}_i,t)\mathbf{v}_i. \tag{2.42}$$

Again, the molecular velocity is decomposed into thermal and advective parts, $\mathbf{v}_i = \mathbf{c}_i + \mathbf{u}(\mathbf{r}_i,t)$, and due to the identity Eq. (2.40) the cross terms vanish, giving the result

$$\mathcal{H}[m_i\mathbf{v}_i] = \sum_i(1 - i\mathbf{k}\cdot\mathbf{r}_i)\mathbf{F}_i - i\mathbf{k}\cdot\sum_i m_i\mathbf{c}_i\mathbf{c}_i - i\mathbf{k}\cdot\sum_i m_i\mathbf{u}(\mathbf{r}_i,t)\mathbf{u}(\mathbf{r}_i,t). \tag{2.43}$$

$\mathbf{F}_i = m_i d\mathbf{v}_i/dt$ is the total force acting on molecule i and is due to conservative interactions with other molecules, denoted $\mathbf{F}_i^c$, and the external force, $\mathbf{F}_i^{\text{ext}}$, that is,

$$\mathbf{F}_i = \mathbf{F}_i^c + \mathbf{F}_i^{\text{ext}}. \tag{2.44}$$

The first term on the right-hand side in Eq. (2.43) then reads

$$\sum_i(1 - i\mathbf{k}\cdot\mathbf{r}_i)\mathbf{F}_i = \sum_i(1 - i\mathbf{k}\cdot\mathbf{r}_i)\mathbf{F}_i^{\text{ext}} - i\mathbf{k}\cdot\sum_i \mathbf{r}_i\mathbf{F}_i^c \tag{2.45}$$

since $\sum_i \mathbf{F}_i^c = \mathbf{0}$ and by using the identity Eq. (2.16). Substitution gives

$$\mathcal{H}[m_i \mathbf{v}_i] = \sum_i (1 - i\mathbf{k} \cdot \mathbf{r}_i) \mathbf{F}_i^{\text{ext}} - i\mathbf{k} \cdot \sum_i m_i \mathbf{u}(\mathbf{r}_i, t) \mathbf{u}(\mathbf{r}_i, t)$$

$$- i\mathbf{k} \cdot \left(\sum_i m_i \mathbf{c}_i \mathbf{c}_i + \sum_i \mathbf{r}_i \mathbf{F}_i^c \right). \tag{2.46}$$

By comparison with Eq. (2.8), one identifies the first term in the Taylor expansions

$$\widetilde{\boldsymbol{\sigma}}(\mathbf{k}, t) = \sum_i (1 - i\mathbf{k} \cdot \mathbf{r}_i) \mathbf{F}_i^{\text{ext}} + \dots \tag{2.47a}$$

$$\widetilde{\rho \mathbf{u}\mathbf{u}}(\mathbf{k}, t) = \sum_i m_i \mathbf{u}(\mathbf{r}_i, t) \mathbf{u}(\mathbf{r}_i, t) + \dots \tag{2.47b}$$

$$\widetilde{\mathbf{P}}(\mathbf{k}, t) = \sum_i m_i \mathbf{c}_i \mathbf{c}_i + \sum_i \mathbf{r}_i \mathbf{F}_i^c + \dots. \tag{2.47c}$$

The momentum flux in real space, $\mathbf{P}$, also has units of force per unit area, that is, pressure, and is therefore referred to as the pressure tensor (and is a rank-2 tensor). From the relations in Eq. (2.47), the Fourier coefficient dynamics is written as

$$\frac{\partial}{\partial t} \widetilde{\rho \mathbf{u}}(\mathbf{k}, t) = \widetilde{\boldsymbol{\sigma}}(\mathbf{k}, t) - i\mathbf{k} \cdot \widetilde{\rho \mathbf{u}\mathbf{u}}(\mathbf{k}, t) - i\mathbf{k} \cdot \widetilde{\mathbf{P}}(\mathbf{k}, t). \tag{2.48}$$

Thus, by simply deriving the different terms entering the balance equation to first order in wavevector we can infer the general balance equation in Fourier space.

It is informative to decompose the pressure tensor into different contributions. For zero wavevector we just showed that

$$\widetilde{\mathbf{P}}(\mathbf{k} = \mathbf{0}, t) = \sum_i m_i \mathbf{c}_i \mathbf{c}_i + \sum_i \mathbf{r}_i \mathbf{F}_i^c. \tag{2.49}$$

The first term is the kinetic part (as it depends only on the thermal velocities) and the second term is the configurational part (as it depends on the positions). The configurational part can be rewritten by assuming pair interactions only, and by Newton's third law we have

$$\sum_i \mathbf{r}_i \mathbf{F}_i^c = \sum_i \mathbf{r}_i \sum_{j \neq i} \mathbf{F}_{ij} = \sum_i \sum_{j > i} \mathbf{r}_{ij} \mathbf{F}_{ij}, \tag{2.50}$$

where $\mathbf{F}_{ij}$ is the conservative force acting on i due to j, and $\mathbf{r}_{ij} = \mathbf{r}_i - \mathbf{r}_j$ is the vector of separation. This means that

$$\widetilde{\mathbf{P}}(\mathbf{k} = \mathbf{0}, t) = \sum_i m_i \mathbf{c}_i \mathbf{c}_i + \sum_i \sum_{j > i} \mathbf{r}_{ij} \mathbf{F}_{ij}. \tag{2.51}$$

In real space this is

$$\mathbf{P}_0(t) = \frac{1}{V} \left[\sum_i m_i \mathbf{c}_i \mathbf{c}_i + \sum_i \sum_{j > i} \mathbf{r}_{ij} \mathbf{F}_{ij} \right], \tag{2.52}$$

where V is the system volume. Equation (2.52) is the famous Irving–Kirkwood [116] expression for the pressure tensor.

The kinetic part of the pressure tensor is the outer product of the thermal velocity and is therefore as previously shown, a rank-2 symmetric tensor. The configurational

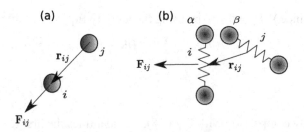

Figure 2.2   Illustrations of the situations where the pressure tensor is symmetric and non-symmetric. (a) Symmetry: the vector of separation $\mathbf{r}_{ij}$ is always parallel with the force, $\mathbf{F}_{ij}$. (b) Non-symmetry: the vector of separation $\mathbf{r}_{ij}$ is, in general, not parallel with the force, $\mathbf{F}_{ij}$.

part, on the other hand, is an outer product between two vectors which are not in general parallel, and is therefore not guaranteed to be symmetric.

We will make the distinction between atomistic systems and molecular systems. The atomistic system models the case where the particles, that is, atoms, groups of atoms or entire molecules, are considered as being simple structureless point masses; see Fig. 2.2(a). In this case the vector of separation $\mathbf{r}_{ij}$ is always parallel to the pair force $\mathbf{F}_{ij}$, and the outer product is symmetric. Hence, the pressure tensor is symmetric. For structured molecules, this is, however, not the case. Figure 2.2(b) illustrates this point with two dumbbell molecules. Here only atoms α in molecule i and β in molecule j will have any significant interaction, and the force on i due to j then acts horizontally. The vector of separation of the two centres of mass is not parallel to the force, and the outer product $\mathbf{r}_{ij}\mathbf{F}_{ij}$ is therefore not symmetric. Note that while the configurational part of the pressure tensor is not, in general, symmetric, the kinetic part of the pressure tensor is.

For molecular systems, one can adopt a purely atomistic definition of the pressure tensor where i and j loop over atoms composing the molecules rather than looping over the molecules. The two formalisms are related by the mass dispersion tensor; the interested reader is referred to [4, 195]. We will proceed using the molecular formalism here.

A non-symmetric tensor can be written as a sum of the symmetric and the antisymmetric parts. For the pressure tensor at zero wavevector this means that

$$\mathbf{P}_0 = \overset{s}{\mathbf{P}}_0 + \overset{a}{\mathbf{P}}_0, \tag{2.53}$$

where the symmetric and antisymmetric parts are, respectively, defined as

$$\overset{s}{\mathbf{P}}_0 = \frac{1}{2}(\mathbf{P}_0 + \mathbf{P}_0^T) \ \text{ and } \ \overset{a}{\mathbf{P}}_0 = \frac{1}{2}(\mathbf{P}_0 - \mathbf{P}_0^T). \tag{2.54}$$

Notice that because the kinetic part of the pressure tensor is symmetric, the thermal velocity will not enter the antisymmetric part of the pressure tensor.

The symmetric part of the pressure tensor is further decomposed into a sum of the diagonal components, the trace, and a traceless part, that is,

$$\overset{s}{\mathbf{P}}_0 = \frac{1}{3}\text{Tr}(\mathbf{P}_0)\mathbf{I} + \overset{os}{\mathbf{P}}_0, \tag{2.55}$$

where $\mathbf{I}$ is the rank-2 identity tensor. Now, the trace defines the equilibrium normal pressure, p_{eq}, and the viscous pressure, Π, which is non-zero for compressible flows, and the symmetric part of the pressure tensor is written up in the final form:

$$\overset{s}{\mathbf{P}}_0 = (p_{eq} + \Pi)\mathbf{I} + \overset{os}{\mathbf{P}}_0 . \tag{2.56}$$

The antisymmetry of $\overset{a}{\mathbf{P}}$ is clear if we write up the tensor component from the definition

$$2\overset{a}{\mathbf{P}}_0 = \begin{bmatrix} 0 & P_{0,xy} - P_{0,yx} & P_{0,xz} - P_{0,zx} \\ P_{0,yx} - P_{0,xy} & 0 & P_{0,yz} - P_{0,zy} \\ P_{0,zx} - P_{0,xz} & P_{0,zy} - P_{0,yz} & 0 \end{bmatrix} . \tag{2.57}$$

By inspection it can be seen that there are three independent components. These can be represented by a pseudo-vector rather than a rank-2 tensor, namely as

$$\overset{ad}{\mathbf{P}}_0 = (\overset{a}{P}_{0,yz}, \overset{a}{P}_{0,zx}, \overset{a}{P}_{0,xy}) . \tag{2.58}$$

From the Irving–Kirkwood definition, Eq. (2.52), we can write the vector components in terms of the forces and positions:

$$\begin{aligned} 2V\overset{ad}{\mathbf{P}}_0 &= \sum_i \left(y_i F^c_{z,i} - z_i F^c_{y,i}, z_i F^c_{x,i} - x_i F^c_{z,i}, x_i F^c_{y,i} - y_i F^c_{x,i} \right) \\ &= \sum_i \mathbf{r}_i \times \mathbf{F}^c_i = \sum_i \sum_{j>i} \mathbf{r}_{ij} \times \mathbf{F}_{ij} . \end{aligned} \tag{2.59}$$

We see that the antisymmetric pressure is due to the torque on i about j, resulting in a change of the orbital angular momentum of the centre of mass of molecule i.

This pressure tensor decomposition is not limited to zero wavevector but can be carried out in general; that is, dropping the subscript, we get

$$\mathbf{P} = (p_{eq} + \Pi)\mathbf{I} + \overset{os}{\mathbf{P}} + \overset{a}{\mathbf{P}} . \tag{2.60}$$

The dynamics of the Fourier coefficients for the linear momentum then take the form

$$\begin{aligned} \frac{\partial \widetilde{\rho \mathbf{u}}}{\partial t} &= \widetilde{\boldsymbol{\sigma}} - i\mathbf{k} \cdot (\widetilde{\rho \mathbf{u}\mathbf{u}}) - i\mathbf{k} \cdot \widetilde{\mathbf{P}} \\ &= \widetilde{\boldsymbol{\sigma}} - i\mathbf{k} \cdot (\widetilde{\rho \mathbf{u}\mathbf{u}}) - i\mathbf{k} \cdot (\widetilde{p}_{eq} + \widetilde{\Pi})\mathbf{I} - i\mathbf{k} \cdot \overset{os}{\widetilde{\mathbf{P}}} - i\mathbf{k} \times \overset{ad}{\widetilde{\mathbf{P}}}, \end{aligned} \tag{2.61}$$

using the identity $i\mathbf{k} \cdot \overset{a}{\widetilde{\mathbf{P}}} = i\mathbf{k} \times \overset{ad}{\widetilde{\mathbf{P}}}$. From Eqs. (2.8a) and (2.8b) we see that Eq. (2.61) is the Fourier transform of the balance equation in real space:

$$\frac{\partial \rho \mathbf{u}}{\partial t} = \boldsymbol{\sigma} - \nabla \cdot (\rho \mathbf{u}\mathbf{u}) - \nabla \cdot ((p_{eq} + \Pi)\mathbf{I} + \overset{os}{\mathbf{P}}) - \nabla \times \overset{ad}{\mathbf{P}} . \tag{2.62}$$

For a system composed of point mass particles the last term is zero, and notice that all terms have the same tensorial character.

Equation (2.62) is known as the conservation form of the linear momentum balance equation. Often one writes the balance equations in a slightly different form using the mass balance equation, Eq. (2.32), and for completeness we derive this here. From the chain rule we have

$$\frac{\partial \rho \mathbf{u}}{\partial t} + \boldsymbol{\nabla} \cdot (\rho \mathbf{u} \mathbf{u}) = \rho \frac{\partial \mathbf{u}}{\partial t} - \mathbf{u}(\boldsymbol{\nabla} \cdot \rho \mathbf{u}) + \rho \mathbf{u} \cdot \boldsymbol{\nabla} \mathbf{u} + \mathbf{u}(\boldsymbol{\nabla} \cdot \rho \mathbf{u})$$

$$= \rho \left(\frac{\partial \mathbf{u}}{\partial t} + \mathbf{u} \cdot \boldsymbol{\nabla} \mathbf{u} \right). \tag{2.63}$$

Note that for the last term we have used the identity $\mathbf{u} \cdot (\boldsymbol{\nabla} \mathbf{u}) = (\mathbf{u} \cdot \boldsymbol{\nabla})\mathbf{u}$. This result can be generalised for any associated field variable, and this defines the so-called material derivative as

$$\frac{\mathrm{D}\phi}{\mathrm{D}t} = \frac{\partial \phi}{\partial t} + \mathbf{u} \cdot \boldsymbol{\nabla}\phi. \tag{2.64}$$

Thus, the linear momentum balance equation can be written as

$$\rho \frac{\mathrm{D}\mathbf{u}}{\mathrm{D}t} = \boldsymbol{\sigma} - \boldsymbol{\nabla} \cdot ((p_{eq} + \Pi)\mathbf{I} + \overset{os}{\mathbf{P}}) - \boldsymbol{\nabla} \times \overset{ad}{\mathbf{P}} \tag{2.65}$$

and is referred to as the convective form.

The balance equation is derived by studying only the small wavevector limit, that is, by application of the $\mathcal{H}$-operator. From this we also obtained the zero wavevector microscopic expression for the pressure tensor which is the Irving–Kirkwood definition. We can derive the wavevector-dependent pressure tensor by application of the $\mathcal{G}$-operator; this can give valuable insight into the multi-scale fluid internal stress relaxations. Now, since $a_i = m_i \mathbf{v}_i$ we have

$$\mathcal{G}[m_i \mathbf{v}_i] = \sum_i \left(m_i \frac{\mathrm{d}\mathbf{v}_i}{\mathrm{d}t} - i\mathbf{k} \cdot (m_i \mathbf{c}_i \mathbf{v}_i) - i\mathbf{k} \cdot (m_i \mathbf{u} \mathbf{v}_i) \right) e^{-i\mathbf{k}\cdot\mathbf{r}_i}$$

$$= \sum_i \left(\mathbf{F}_i - i\mathbf{k} \cdot m_i \mathbf{c}_i \mathbf{c}_i - i\mathbf{k} \cdot m_i \mathbf{u}(\mathbf{r}_i, t)\mathbf{u}(\mathbf{r}_i, t) \right) e^{-i\mathbf{k}\cdot\mathbf{r}_i}.$$

$$\tag{2.66}$$

Again, $\mathbf{F}_i$ is the total force acting on molecule i due to interaction with other molecules, denoted $\mathbf{F}_i^c$, and $\mathbf{F}^{\mathrm{ext}}$, the external force. Thus,

$$\sum_i \mathbf{F}^{\mathrm{tot}} e^{-i\mathbf{k}\cdot\mathbf{r}_i} = \sum_i \mathbf{F}_i^c e^{-i\mathbf{k}\cdot\mathbf{r}_i} + \sum_i \mathbf{F}^{\mathrm{ext}} e^{-i\mathbf{k}\cdot\mathbf{r}_i}. \tag{2.67}$$

Assuming pairwise interactions only and by application of Newton's third law, $\mathbf{F}_{ij} = -\mathbf{F}_{ji}$,

$$\sum_i \mathbf{F}_i^c e^{-i\mathbf{k}\cdot\mathbf{r}_i} = \sum_i e^{-i\mathbf{k}\cdot\mathbf{r}_i} \sum_{j \neq i} \mathbf{F}_{ij}$$

$$= \sum_i \sum_{j > i} \left(e^{-i\mathbf{k}\cdot\mathbf{r}_i} - e^{-i\mathbf{k}\cdot\mathbf{r}_j} \right) \mathbf{F}_{ij}$$

$$= -\sum_i \sum_{j > i} \mathbf{F}_{ij} (e^{i\mathbf{k}\cdot\mathbf{r}_{ij}} - 1) e^{-i\mathbf{k}\cdot\mathbf{r}_i}. \tag{2.68}$$

Moreover, for non-zero wavevector the force can be written as

$$\mathbf{F}_{ij} = \frac{\mathbf{F}_{ij} \cdot (i\mathbf{k} \cdot \mathbf{r}_{ij})}{i\mathbf{k} \cdot \mathbf{r}_{ij}} = \frac{i\mathbf{k} \cdot (\mathbf{r}_{ij} \mathbf{F}_{ij})}{i\mathbf{k} \cdot \mathbf{r}_{ij}} \quad (\mathbf{k} \neq \mathbf{0}) \tag{2.69}$$

using the relation given in Eq. (2.16) and we have

$$\sum_i \mathbf{F}_i^c e^{-i\mathbf{k}\cdot\mathbf{r}_i} = -i\mathbf{k}\cdot\sum_i\sum_{j>i}\mathbf{r}_{ij}\mathbf{F}_{ij}\frac{e^{i\mathbf{k}\cdot\mathbf{r}_{ij}}-1}{i\mathbf{k}\cdot\mathbf{r}_{ij}}e^{-i\mathbf{k}\cdot\mathbf{r}_i}. \tag{2.70}$$

The dynamics are given by substitution into Eq. (2.66):

$$\mathcal{G}[m_i\mathbf{v}_i] = \sum_i\mathbf{F}_i^{\text{ext}}e^{-i\mathbf{k}\cdot\mathbf{r}_i} - i\mathbf{k}\cdot\sum_i m_i\mathbf{u}(\mathbf{r}_i,t)\mathbf{u}(\mathbf{r}_i,t)e^{-i\mathbf{k}\cdot\mathbf{r}_i}$$

$$- i\mathbf{k}\cdot\sum_i\left(m_i\mathbf{c}_i\mathbf{c}_i + \sum_{j>i}\mathbf{r}_{ij}\mathbf{F}_{ij}\frac{e^{i\mathbf{k}\cdot\mathbf{r}_{ij}}-1}{i\mathbf{k}\cdot\mathbf{r}_{ij}}\right)e^{-i\mathbf{k}\cdot\mathbf{r}_i}, \tag{2.71}$$

and each term in Eq. (2.7) is readily identified. In particular, the wavevector-dependent pressure tensor is

$$\widetilde{\mathbf{P}}(\mathbf{k},t) = \sum_i\left(m_i\mathbf{c}_i\mathbf{c}_i + \sum_{j>i}\mathbf{r}_{ij}\mathbf{F}_{ij}\frac{e^{i\mathbf{k}\cdot\mathbf{r}_{ij}}-1}{i\mathbf{k}\cdot\mathbf{r}_{ij}}\right)e^{-i\mathbf{k}\cdot\mathbf{r}_i}. \tag{2.72}$$

The Irving–Kirkwood tensor is recaptured in the zero wavevector limit

$$\lim_{\mathbf{k}\to 0}\frac{e^{i\mathbf{k}\cdot\mathbf{r}_{ij}}-1}{i\mathbf{k}\cdot\mathbf{r}_{ij}} = 1, \tag{2.73}$$

and the pressure tensor in this limit is

$$\lim_{\mathbf{k}\to 0}\widetilde{\mathbf{P}}(\mathbf{k},t) = \sum_i m_i\mathbf{c}_i\mathbf{c}_i + \sum_i\sum_{j>i}\mathbf{r}_{ij}\mathbf{F}_{ij}, \tag{2.74}$$

in agreement with Eq. (2.52). See also Evans and Morriss [66] and Todd and Daivis [195]. It is worth noting that the wavevector-dependent pressure tensor can also be computed directly from the momentum balance equation, assuming zero advection; see Section 2.4, 'Further Explorations'.

2.2.3 Thermal Kinetic Energy Balance

As mentioned in the introduction to this section, the classical treatment deals with the mass, linear momentum, and total energy balance equations which form the dynamical set of equations. We have seen how to apply the microscopic hydrodynamic operators in order to derive the first two equations. Rather than deriving the balance equation for the total energy, we here follow Alley and Alder [6] and derive the balance equation for the thermal kinetic energy density, $\rho\varepsilon$, This immediately leads to the equation for the kinetic temperature that we will use in our further explorations. Furthermore, in order to avoid fatigue we derive the balance equation for the zero-flow situation and simply list the more general flow case and refer to the existing literature.

We do not assign the thermal kinetic energy density a symbol, but write it in terms of the product of density ρ and the associated field variable, ε, the thermal kinetic energy per unit mass.

The thermal kinetic energy density is microscopically defined from the molecular thermal velocities

$$\rho(\mathbf{r},t)\varepsilon(\mathbf{r},t) = \sum_i \varepsilon_i \delta(\mathbf{r}-\mathbf{r}_i) = \frac{1}{2}\sum_i m_i c_i^2 \delta(\mathbf{r}-\mathbf{r}_i), \tag{2.75}$$

where $c_i^2 = \mathbf{c}_i \cdot \mathbf{c}_i$ is the usual dot product. As we have stated, we limit ourselves to the case $\mathbf{u} = \mathbf{0}$ at all times and in every point. This instantaneous zero-flow assumption implies that $\mathbf{F}_i = \mathbf{F}_i^c$, that is, only conservative forces act on the molecules. In this case the $\mathcal{H}$-operator reads

$$\mathcal{H}[\varepsilon_i] = \sum_i (1 - i\mathbf{k}\cdot\mathbf{r}_i)\frac{d\varepsilon_i}{dt} - i\mathbf{k}\cdot\sum_i \mathbf{c}_i \varepsilon_i. \tag{2.76}$$

First term on the right-hand side is (leaving out the summation)

$$\begin{aligned}
(1 - i\mathbf{k}\cdot\mathbf{r}_i)\frac{d\varepsilon_i}{dt} &= (1 - i\mathbf{k}\cdot\mathbf{r}_i)\frac{m_i}{2}\frac{d}{dt}(\mathbf{c}_i \cdot \mathbf{c}_i)\\
&= (1 - i\mathbf{k}\cdot\mathbf{r}_i)m_i\mathbf{c}_i\cdot\frac{d\mathbf{c}_i}{dt}\\
&= (1 - i\mathbf{k}\cdot\mathbf{r}_i)(\mathbf{c}_i\cdot\mathbf{F}_i)\\
&= \mathbf{c}_i\cdot\mathbf{F}_i - i\mathbf{k}\cdot(\mathbf{r}_i\mathbf{F}_i)\cdot\mathbf{c}_i.
\end{aligned} \tag{2.77}$$

For the second term in Eq. (2.76) we have

$$\mathbf{c}_i\varepsilon_i = \frac{1}{2}m_i c_i^2 \mathbf{c}_i = \frac{1}{2}m_i\mathbf{c}_i\mathbf{c}_i\cdot\mathbf{c}_i. \tag{2.78}$$

Collecting the terms, the microscopic hydrodynamic operator reads (this time with the summations)

$$\mathcal{H}[\varepsilon_i] = \sum_i \mathbf{c}_i\cdot\mathbf{F}_i - i\mathbf{k}\cdot\sum_i\left(\frac{1}{2}m_i\mathbf{c}_i\mathbf{c}_i + \mathbf{r}_i\mathbf{F}_i\right)\cdot\mathbf{c}_i. \tag{2.79}$$

The first term is wavevector independent, hence, the thermal kinetic energy density is not a conserved quantity as we expect. The second term in the $\mathcal{H}$-operator is the diffusive process. The terms in the balance equation are then identified as

$$\widetilde{\sigma}_\varepsilon(\mathbf{k},t) = \sum_i \mathbf{c}_i\cdot\mathbf{F}_i + \ldots \tag{2.80a}$$

$$\widetilde{\mathbf{J}}^\varepsilon(\mathbf{k},t) = \sum_i\left(\frac{1}{2}m_i\mathbf{c}_i\mathbf{c}_i + \mathbf{r}_i\mathbf{F}_i\right)\cdot\mathbf{c}_i + \ldots, \tag{2.80b}$$

and the balance equation for the Fourier coefficients is

$$\frac{\partial\widetilde{\rho\varepsilon}}{\partial t} = \widetilde{\sigma}_\varepsilon - i\mathbf{k}\cdot\widetilde{\mathbf{J}}^\varepsilon, \tag{2.81}$$

which is the Fourier transform of

$$\frac{\partial\rho\varepsilon}{\partial t} = \sigma_\varepsilon - \mathbf{\nabla}\cdot\mathbf{J}^\varepsilon. \tag{2.82}$$

$\mathbf{J}^\varepsilon$ is here referred to as the thermal kinetic energy flux. Let us address the production term σ_ε. First, we assume (i) that the system is locally in thermodynamic equilibrium,

(ii) that the local density change is large compared to the change in heat; this is the local adiabatic approximation, and (iii) the kinetic temperature is the same as the thermodynamic temperature, which is strictly true only at equilibrium.

If the system is in local thermodynamic equilibrium, the system state can be described by the entropy function $S = S(T,V,N)$, where T is the temperature, V the volume, and N the number of particles. If we keep the local number of particles fixed, then the entropy is dependent only on temperature and number density $n = N/V$, that is, $S = S(T,n)$. The change in the (local) entropy is therefore to first order

$$dS = \left(\frac{\partial S}{\partial T}\right)_n dT + \left(\frac{\partial S}{\partial n}\right)_T dn. \tag{2.83}$$

where subscripts indicate that n and T are held fixed. Using the chain rule and the Maxwell relation $(\partial S/\partial V)_T = (\partial p/\partial T)_V$, we obtain

$$\begin{aligned} dS &= \left(\frac{\partial S}{\partial T}\right)_n dT - \frac{V}{n}\left(\frac{\partial p}{\partial T}\right)_V dn \\ &= \frac{mNc_V}{T}dT - \frac{V\beta_V}{\rho}d\rho, \end{aligned} \tag{2.84}$$

where c_V is the specific heat capacity at constant volume, β_V is the thermal pressure coefficient, $\beta_V = (\partial p/\partial T)_\rho$, and $\rho = mn$ as we consider a single-component fluid. Now, the entropy and heat Q are related through $dS = dQ/T$; thus, we can write the heat density as

$$\frac{dQ}{V} = \rho c_V dT - \frac{T\beta_V}{\rho}d\rho. \tag{2.85}$$

By invoking the local adiabatic approximation, $dQ = 0$, we arrive at

$$dT = \frac{T\beta_V}{\rho^2 c_V}d\rho. \tag{2.86}$$

In thermodynamic equilibrium the thermal kinetic energy and kinetic temperature are related through

$$c_V\rho T = \rho\varepsilon, \tag{2.87}$$

and therefore

$$d\rho\varepsilon = \frac{T\beta_V}{\rho}d\rho. \tag{2.88}$$

The local rate of change (i.e., the production term) is therefore

$$\sigma_\varepsilon = \frac{T\beta_V}{\rho}\frac{\partial \rho}{\partial t}. \tag{2.89}$$

Notice that the thermal kinetic energy then couples to the mass density. Substituting Eq. (2.89) into Eq. (2.82), we arrive at the balance equation for the thermal kinetic energy in the zero flow case

$$\frac{\partial \rho\varepsilon}{\partial t} = \frac{T\beta_V}{\rho}\frac{\partial \rho}{\partial t} - \nabla \cdot \mathbf{J}^\varepsilon. \tag{2.90}$$

The zero-flow balance equation for the kinetic temperature is readily obtained by inserting Eq. (2.87) into Eq. (2.90).

In the general case an additional advection term and a viscous heating term arise. In fact, we will see two examples of this in the last chapter, and it is relevant to write the general result

$$c_V \frac{\partial \rho T}{\partial t} = \frac{T \beta_V}{\rho} \frac{\partial \rho}{\partial t} - c_V \nabla \cdot (\rho \mathbf{u} T) - \nabla \cdot \mathbf{J}^\varepsilon - \mathbf{P}^T : \nabla \mathbf{u}. \tag{2.91}$$

The viscous heating term is given by a double contraction of the pressure tensor transposed, $\mathbf{P}^T$, and the outer product, $\nabla \mathbf{u}$. In general, the double contraction is defined as

$$\mathbf{A} : \mathbf{B} = \sum_i \sum_j A_{ij} B_{ji}. \tag{2.92}$$

For more details the interested reader is referred to Todd and Daivis [200] and Evans and Morriss [66].

2.3 More Examples

The microscopic hydrodynamic operators offer a general framework to derive the balance equation for any dynamical variable. In this section this will be shown for the balance equations for the angular momenta and the polarisation.

2.3.1 Angular Momenta Balances

We make the distinction between the orbital angular momentum density, $\rho \mathbf{L}$, and the spin angular momentum, $\rho \mathbf{S}$. The former is the angular momentum of the fluid with respect to a given origin, and the latter the angular momentum with respect to the fluid element centre of mass. As for the thermal kinetic energy, we will not assign symbols for the two angular momentum densities, but simply write them in terms of the density and the associated field. Recall that in the limit of zero wavevector the total angular momentum is a constant of motion, that is, it is a conserved quantity.

We start from the orbital angular momentum density, $\rho \mathbf{L} = \rho(\mathbf{r}, t) \mathbf{L}(\mathbf{r}, t)$. The molecular definition is

$$\rho(\mathbf{r}, t) \mathbf{L}(\mathbf{r}, t) = \sum_i \mathbf{L}_i \delta(\mathbf{r} - \mathbf{r}_i) = \sum_i (\mathbf{r}_i \times \mathbf{p}_i) \delta(\mathbf{r} - \mathbf{r}_i), \tag{2.93}$$

where $\mathbf{L}_i$ is the molecular angular momentum, and $\mathbf{p}_i$ is the centre-of-mass momentum. Note, the orbital angular momentum is defined from the cross product of two vectors and is therefore a pseudo-vector.

In the following we let the external force be zero. $\mathcal{H}$ acting on the molecular variable yields

$$\mathcal{H}[\mathbf{L}_i] = \sum_i (1 - i\mathbf{k} \cdot \mathbf{r}_i) \frac{d\mathbf{L}_i}{dt} - i\mathbf{k} \cdot \sum_i \mathbf{c}_i \mathbf{L}_i - i\mathbf{k} \cdot \sum_i \mathbf{u}(\mathbf{r}_i, t) \mathbf{L}_i. \tag{2.94}$$

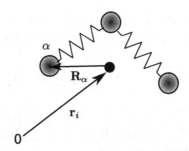

Figure 2.3 Illustration of the vectors $\mathbf{R}_\alpha$ and $\mathbf{r}_i$ entering Eq. (2.99).

The first term on the right-hand side is (again, excluding the summation over molecules)

$$
\begin{aligned}
(1 - i\mathbf{k} \cdot \mathbf{r}_i)\frac{d\mathbf{L}_i}{dt} &= (1 - i\mathbf{k} \cdot \mathbf{r}_i)\frac{d}{dt}(\mathbf{r}_i \times \mathbf{p}_i) \\
&= (1 - i\mathbf{k} \cdot \mathbf{r}_i)\left(\frac{d\mathbf{r}_i}{dt} \times \mathbf{p}_i + \mathbf{r}_i \times \frac{d\mathbf{p}_i}{dt}\right) \\
&= \mathbf{N}_i - i\mathbf{k} \cdot \mathbf{r}_i \mathbf{N}_i,
\end{aligned}
\tag{2.95}
$$

where $\mathbf{N}_i = \mathbf{r}_i \times \mathbf{F}_i^c$ is the torque on molecule i. The $\mathcal{H}$-operator can then be written as

$$
\mathcal{H}[\mathbf{L}_i] = \mathbf{N} - i\mathbf{k} \cdot \sum_i (\mathbf{c}_i \mathbf{L}_i - \mathbf{r}_i \mathbf{N}_i) - i\mathbf{k} \cdot \sum_i \mathbf{u}(\mathbf{r}_i, t)\mathbf{L}_i,
\tag{2.96}
$$

where, by virtue of Eq. (2.59), we have that the total torque is

$$
\mathbf{N} = \sum_i \mathbf{r}_i \times \mathbf{F}_i^c = 2V \overset{ad}{\mathbf{P}_0}.
\tag{2.97}
$$

From this we can infer the orbital angular momentum balance equation for the Fourier coefficients,

$$
\frac{\partial}{\partial t}\widetilde{\rho\mathbf{L}}(\mathbf{k}, t) = 2\overset{ad}{\widetilde{\mathbf{P}}} - i\mathbf{k} \cdot \widetilde{\rho\mathbf{u}\mathbf{L}} - i\mathbf{k} \cdot \widetilde{\mathbf{J}}^L,
\tag{2.98}
$$

where $\mathbf{J}^L$ is the orbital angular momentum flux tensor; the zero wavevector limit is given in Eq. (2.96). We recognise that $\rho\mathbf{L}$ is not a conserved quantity.

The spin angular momentum density, $\rho\mathbf{S} = \rho(\mathbf{r}, t)\mathbf{S}(\mathbf{r}, t)$, is defined from the molecular angular momenta by

$$
\rho(\mathbf{r}, t)\mathbf{S}(\mathbf{r}, t) = \sum_i \mathbf{S}_i \delta(\mathbf{r} - \mathbf{r}_i) = \sum_i \left(\sum_{\alpha \in i} \mathbf{R}_\alpha \times \mathbf{p}_\alpha\right) \delta(\mathbf{r} - \mathbf{r}_i).
\tag{2.99}
$$

The index α denotes atom or particle α in molecule i, $\mathbf{R}_\alpha$ is the vector from the molecular centre of mass to atom α; see Fig. 2.3. $\mathbf{S}_i = \sum_{\alpha \in i} \mathbf{R}_\alpha \times \mathbf{p}_\alpha$ is then the molecular spin angular momentum (i.e., the rotation around the centre of mass), and $\mathbf{p}_\alpha$ is the momentum of atom α. As the case for the orbital angular momentum, we see from the definition that the spin angular momentum is a pseudo-vector.

Letting the $\mathcal{H}$-operator act on $\mathbf{S}_i$, one gets the dynamics in the small wavevector limit (here leaving out a few details in the derivation),

$$\mathcal{H}[\mathbf{S}_i] = \mathbf{M} - i\mathbf{k} \cdot \left(\sum_i \mathbf{c}_i \mathbf{S}_i - \sum_i \mathbf{r}_i \mathbf{M}_i \right) - i\mathbf{k} \cdot \sum_i \mathbf{u}(\mathbf{r}_i, t) \mathbf{S}_i, \qquad (2.100)$$

where $\mathbf{M}_i = \sum_{\alpha \in i} \mathbf{R}_\alpha \times \mathbf{F}_\alpha^C$ is the total torque on i with respect to the centre of mass and $\mathbf{M} = \sum_i \mathbf{M}_i$. From the $\mathcal{H}$-operator we readily see that the spin angular momentum is not conserved as expected.

The second term on the right-hand side of Eq. (2.100) is the Irving–Kirkwood spin angular momentum flux tensor at zero wavevector, that is,

$$\begin{aligned} \tilde{\mathbf{Q}}(\mathbf{k} = \mathbf{0}, t) &= \sum_i \mathbf{c}_i \mathbf{S}_i + \sum_i \mathbf{r}_i \mathbf{M}_i \\ &= \sum_i \mathbf{c}_i \mathbf{S}_i + \sum_i \sum_{j>i} \mathbf{r}_{ij} \mathbf{M}_{ij}. \end{aligned} \qquad (2.101)$$

$\mathbf{M}_{ij}$ is the torque on molecule i due to molecule j. From the first-order wavevector dynamics, Eq. (2.100), we get the general balance equation for the spin angular momentum,

$$\frac{\partial}{\partial t} \widetilde{\rho \mathbf{S}}(\mathbf{k}, t) = \tilde{\mathbf{M}} - i\mathbf{k} \cdot \widetilde{\rho \mathbf{u} \mathbf{S}} - i\mathbf{k} \cdot \tilde{\mathbf{Q}}. \qquad (2.102)$$

This equation can be reformulated. We recall that the total angular momentum is a conserved quantity, which implies that for zero wavevector

$$\frac{\partial}{\partial t} (\widetilde{\rho \mathbf{L}} + \widetilde{\rho \mathbf{S}}) = \mathbf{0} \quad (\mathbf{k} = \mathbf{0}), \qquad (2.103)$$

that is, $\mathbf{M} + \mathbf{N} = \mathbf{0}$ according to Eqs.(2.98) and (2.102), and therefore,

$$\mathbf{M} = -2V \overset{ad}{\mathbf{P}}_0. \qquad (2.104)$$

Moreover, it can be seen that the spin angular momentum flux tensor is not symmetric; and as for the pressure tensor, it can be decomposed into the trace, traceless symmetric, and antisymmetric parts, that is, in general we have

$$\mathbf{Q} = Q\mathbf{I} + \overset{os}{\mathbf{Q}} + \overset{a}{\mathbf{Q}}. \qquad (2.105)$$

Writing the antisymmetric part as the vector $\overset{ad}{\mathbf{Q}}$, the balance equation for the spin angular momentum can be written as

$$\frac{\partial}{\partial t} \widetilde{\rho \mathbf{S}}(\mathbf{k}, t) = -2 \overset{ad}{\tilde{\mathbf{P}}} - i\mathbf{k} \cdot (\widetilde{\rho \mathbf{u} \mathbf{S}}) - i\mathbf{k} \cdot \tilde{Q}\mathbf{I} - i\mathbf{k} \cdot \overset{os}{\tilde{\mathbf{Q}}} - i\mathbf{k} \times \overset{ad}{\tilde{\mathbf{Q}}} \qquad (2.106)$$

in the absence of external driving forces. This result is the Fourier transformation of the real space balance equation,

$$\frac{\partial}{\partial t} \rho \mathbf{S}(\mathbf{r}, t) = -2 \overset{ad}{\mathbf{P}} - \nabla \cdot (\rho \mathbf{u} \mathbf{S}) - \nabla \cdot (Q\mathbf{I} + \overset{os}{\mathbf{Q}}) - \nabla \times \overset{ad}{\mathbf{Q}}. \qquad (2.107)$$

Since $\overset{ad}{\mathbf{Q}}$ is a vector,[2] we have from Eq. (2.5) that the requirement of the same tensorial character is fulfilled.

More can be done: the spin angular momentum can be written in terms of the spin angular velocity $\mathbf{\Omega}$,

$$\rho(\mathbf{r},t)\mathbf{S}(\mathbf{r},t) = \rho(\mathbf{r},t)\mathbf{\Theta}(\mathbf{r},t) \cdot \mathbf{\Omega}(\mathbf{r},t). \tag{2.108}$$

The inertia tensor per unit mass, $\mathbf{\Theta}$, is the local average molecular inertia. For a single-component system, this is then simply $\mathbf{\Theta} = \mathbf{\Theta}_{\text{mol}}$, where

$$\mathbf{\Theta}_{\text{mol}} = \frac{1}{m_i} \sum_{\alpha \in i} m_\alpha (\mathbf{R}_\alpha^2 \mathbf{I} - \mathbf{R}_\alpha \mathbf{R}_\alpha), \tag{2.109}$$

where $m_i = \sum_{\alpha \in i} m_\alpha$ is the molecule mass. It is clear from Eq. (2.109) that the molecular moment of inertia tensor is real and symmetric. This means that there exists an orthogonal matrix $\mathbf{T}$ such that $\mathbf{\Theta}_P = \mathbf{T}^{-1} \cdot \mathbf{\Theta}_{\text{mol}} \cdot \mathbf{T}$, where $\mathbf{\Theta}_P$ is the moment of inertia diagonal tensor in the principal coordinate system, that is, in the molecular rotating coordinate system spanned by the eigenvectors of $\mathbf{\Theta}_{\text{mol}}$. Since $\mathbf{T}$ is an orthogonal matrix, its inverse equals its transpose, $\mathbf{T}^{-1} = \mathbf{T}^T$, and it can be shown (e.g. via direct computation) that for any vector $\mathbf{a}$ we have the identity $\mathbf{T} \cdot \mathbf{a} = \mathbf{a} \cdot \mathbf{T}^{-1}$. We can then conclude

$$\begin{aligned}
\mathbf{\Theta}_{\text{mol}} \cdot \mathbf{\Omega} &= \mathbf{T} \cdot \mathbf{T}^{-1} \cdot \mathbf{\Theta}_{\text{mol}} \cdot \mathbf{T} \cdot \mathbf{T}^{-1} \cdot \mathbf{\Omega} \\
&= \mathbf{T} \cdot \mathbf{\Theta}_P \cdot \mathbf{T}^{-1} \cdot \mathbf{\Omega} \\
&= \mathbf{T} \cdot (\mathbf{\Theta}_P \cdot \mathbf{\Omega}) \cdot \mathbf{T} \\
&= \mathbf{\Theta}_P \cdot \mathbf{\Omega}.
\end{aligned} \tag{2.110}$$

For rigid molecules, the molecular principal moment of inertia $\mathbf{\Theta}_P$ is constant. If we assume homogeneity and isotropy, the molecular inertia is a scalar quantity and can therefore be expressed through a scalar which is the average of the principal molecular inertia tensor,

$$\Theta = \frac{1}{3} \text{Tr}\mathbf{\Theta}_P. \tag{2.111}$$

In this case, the spin angular momentum density takes a particularly simple form,

$$\rho\mathbf{S} = \rho(\mathbf{r},t)\Theta\mathbf{\Omega}(\mathbf{r},t), \tag{2.112}$$

and the balance equation, Eq. (2.107), is then

$$\Theta\left(\frac{\partial\rho\mathbf{\Omega}}{\partial t} + \mathbf{\nabla}\cdot(\rho\mathbf{u}\mathbf{\Omega})\right) = -2\overset{ad}{\mathbf{P}} - \mathbf{\nabla}\cdot(Q + \overset{os}{\mathbf{Q}}) - \mathbf{\nabla}\times\overset{ad}{\mathbf{Q}}. \tag{2.113}$$

Notice that in the balance equations for the spin angular momentum and the linear momentum, the antisymmetric part of the pressure appears. We will later see that the constitutive model for $\overset{ad}{\mathbf{P}}$ involves both the velocity, $\mathbf{u}$, and the spin angular velocity, $\mathbf{\Omega}$, hence these two associated field variables couple, as was also discussed in the

---

[2] I will leave it up to the reader to verify this.

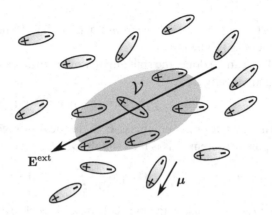

Figure 2.4 Illustration of the polarisation of fluid element $\mathcal{V}$ due to application of an external electric field, $\mathbf{E}^{ext}$. $\boldsymbol{\mu}$ represents the molecular dipole vector. Due to thermal fluctuations, the molecular dipoles are far from perfectly aligned with the field, and the alignment illustrated here is exaggerated.

introduction. This extends the hydrodynamic description. Naturally, we will not expect this coupling to be relevant on the macroscopic length scale; we return to this point in Chapter 4.

2.3.2 Polarisation Balance

Consider a fluid element $\mathcal{V}$ composed of molecules with permanent dipoles. In the presence of an electric field, the molecular dipoles will partially align with the field. The resulting charge on the element surface is a bound charge in that it is bounded by the molecular structure (and therefore the fluid element) and cannot freely move in response to the field. The molecular alignment will induce two opposite non-zero net charges on the fluid element in alignment with the field, and the element becomes polarised; see the illustration in Fig. 2.4.

Of course, not all charge is bounded; for example, ions in an electrolyte solution can perform translational motion in response to a field, and electrons can jump from molecule to molecule. Charge which is not bounded is called free charge [86] and the charge density, ρ_q, can be written as

$$\rho_q = \rho_f + \rho_b, \qquad (2.114)$$

where ρ_f and ρ_b are the free and bound charge densities, respectively.

Now we return to the bound charges. We define the polarisation $\mathbf{P}$, or dipole moment density, at a given point from the bound charge density,

$$\nabla \cdot \mathbf{P} = -\rho_b. \qquad (2.115)$$

This definition can be thought of as a Gauss law for the bound charge. As we see later, Eq. (2.115) gives a direct way to calculate the polarisation in simulations; however, it is not very helpful unless the full set of Maxwell equations is invoked [52, 53].

Alternatively and with some approximations, the $\mathcal{H}$-operator can be applied, giving the balance equation for $\mathbf{P}$, and we pursue this approach here.

We ignore induced polarisation effects and consider only polarisation due to molecular alignment. Also, we will initially not include the presence of an external electric field, but will add this later. Let the dipole moment of molecule i be $\boldsymbol{\mu}_i$ and the polarisation density be defined microscopically as [90],

$$\mathbf{P}(\mathbf{r},t) = \rho(\mathbf{r},t)\mathbf{p}(\mathbf{r},t) = \sum_i \boldsymbol{\mu}_i(t)\delta(\mathbf{r}-\mathbf{r}_i), \tag{2.116}$$

where the associated field $\mathbf{p}$ is the dipole momentum per unit mass. Notice that this definition is only the second-order term in the multipole expansion of $\mathbf{P}$ and will fail at large wavevectors; we return to this in Chapter 4. The $\mathcal{H}$-operator acting on $\boldsymbol{\mu}_i$ is

$$\mathcal{H}[\boldsymbol{\mu}_i] = \sum_i (1 - i\mathbf{k}\cdot\mathbf{r}_i)\frac{d\boldsymbol{\mu}_i}{dt} - i\mathbf{k}\cdot\sum_i \mathbf{c}_i\boldsymbol{\mu}_i - i\mathbf{k}\cdot\sum_i \mathbf{u}(\mathbf{r}_i,t)\boldsymbol{\mu}_i. \tag{2.117}$$

In the principal coordinate system (rotating molecular reference frame), $\boldsymbol{\mu}_i$ is constant since the dipole moment is permanent. In the fixed reference frame, Euler's rotation equation for $\boldsymbol{\mu}_i$ gives

$$\frac{d\boldsymbol{\mu}_i}{dt} = \boldsymbol{\Omega}_i \times \boldsymbol{\mu}_i, \tag{2.118}$$

where $\boldsymbol{\Omega}_i$ is the dipole angular velocity. Substitution in Eq. (2.117) and rearranging the dynamics in the small wavevector regime is governed by

$$\mathcal{H}[\boldsymbol{\mu}_i] = \sum_i \boldsymbol{\Omega}_i \times \boldsymbol{\mu}_i - i\mathbf{k}\cdot\sum_i \mathbf{u}(\mathbf{r}_i,t)\boldsymbol{\mu}_i - i\mathbf{k}\cdot\sum_i (\mathbf{c}_i\boldsymbol{\mu}_i + \mathbf{r}_i(\boldsymbol{\Omega}_i \times \boldsymbol{\mu}_i)). \tag{2.119}$$

The first term on the right-hand side shows that $\mathbf{P}$ is not conserved.

In order to interpret the production term macroscopically, note first that the cross product between the spin angular velocity and the polarisation in real space is

$$\boldsymbol{\Omega} \times \rho\mathbf{p} = \boldsymbol{\Omega}(\mathbf{r},t) \times \sum_i \boldsymbol{\mu}_i(t)\delta(\mathbf{r}-\mathbf{r}_i)$$
$$= \sum_i \boldsymbol{\Omega}(\mathbf{r},t) \times \boldsymbol{\mu}_i(t)\delta(\mathbf{r}-\mathbf{r}_i) \tag{2.120}$$

due to the distributive properties of the cross product. Ignoring density fluctuations, we can write the spin angular velocity at the molecular point $\mathbf{r}_i$ as $\boldsymbol{\Omega}(\mathbf{r}_i,t) \approx \boldsymbol{\Omega}_i(t)$. Then, from Fourier transforming Eq. (2.120), we get

$$\mathcal{F}[\boldsymbol{\Omega} \times \rho\mathbf{p}] = \sum_i \boldsymbol{\Omega}(\mathbf{r}_i,t) \times \boldsymbol{\mu}_i(t)e^{-i\mathbf{k}\cdot\mathbf{r}_i} \approx \sum_i \boldsymbol{\Omega}_i(t) \times \boldsymbol{\mu}_i(t) \tag{2.121}$$

in the limit of small wavevectors. We now see that the production term found from the $\mathcal{H}$-operator is a coupling between the polarisation density (dielectrics) and the spin angular velocity (mechanics); the term is non-linear and only zero if the polarisation follows the local fluid angular velocity.

From Eq. (2.119) one can also immediately define the Irving–Kirkwood dipole flux tensor as

$$\widetilde{\mathbf{R}}(\mathbf{k}=\mathbf{0},t) = \sum_i \mathbf{c}_i\boldsymbol{\mu}_i + \mathbf{r}_i(\boldsymbol{\Omega}_i \times \boldsymbol{\mu}_i), \tag{2.122}$$

and we see that this tensor is not, in general, symmetric. Again, the tensor $\mathbf{R}$ can be composed into trace, traceless symmetric, and antisymmetric parts; however, for brevity we write the balance equation in a more compact form,

$$\frac{\partial}{\partial t}\widetilde{\rho \mathbf{p}}(\mathbf{k},t) = \widetilde{\mathbf{\Omega} \times \rho \mathbf{p}} - i\mathbf{k} \cdot (\widetilde{\rho \mathbf{u} \mathbf{p}}) - i\mathbf{k} \cdot \widetilde{\mathbf{R}}. \tag{2.123}$$

This is the Fourier transform of the real space balance equation,

$$\frac{\partial \rho \mathbf{p}}{\partial t} = \mathbf{\sigma} + \mathbf{\Omega} \times \rho \mathbf{p} - \mathbf{\nabla} \cdot (\rho \mathbf{u} \mathbf{p}) - \mathbf{\nabla} \cdot \mathbf{R}, \tag{2.124}$$

when allowing for a production term.

The last term in this derivation indicates that there exists a diffusive process which tends to remove gradients in the polarisation itself. While this process was described by, for example, Dahler and Scriven [46], in 1963 it is often ignored. However, it must be included in the multiscale description, as we will see later.

In Fig. 1.4 the polarisation was plotted as a function of time for water after application of an external electric field, $\mathbf{E}^{\mathrm{ext}}$. The response is modelled via the Debye equation; this reads in a more general differential form as

$$\frac{\mathrm{d}\rho \mathbf{p}}{\mathrm{d}t} = \frac{1}{\tau_D} \left(\varepsilon_0 \chi_e \mathbf{E}^{\mathrm{ext}} - \rho \mathbf{p} \right). \tag{2.125}$$

This linear differential equation is easily solved once the initial condition is specified. The Debye equation can also be formulated in terms of the local field $\mathbf{E}$, that is, the field due to the external electric field (if this is applied) and the resulting screening field. Even in the case where the external field is zero, there will still be a local thermally fluctuating field arising from the bound charges. For our purposes, Eq. (2.125) is used. Thus, if we let the production term be written by Eq. (2.125), we arrive at the final form for the polarisation balance equation,

$$\frac{\partial \rho \mathbf{p}}{\partial t} = \frac{1}{\tau_D} \left(\varepsilon_0 \chi_e \mathbf{E}^{\mathrm{ext}} - \rho \mathbf{p} \right) + \mathbf{\Omega} \times \rho \mathbf{p} - \mathbf{\nabla} \cdot (\rho \mathbf{u} \mathbf{p}) - \mathbf{\nabla} \cdot \mathbf{R}. \tag{2.126}$$

2.4 Further Explorations

1. Let $\mathbf{a} = (a_x, a_y, a_z), \mathbf{b} = (b_x, b_y, b_z)$, and $\mathbf{c} = (c_x, c_y, c_z)$. From direct calculus, verify the identity, Eq. (2.16),

$$\mathbf{a}(\mathbf{b} \cdot \mathbf{c}) = \mathbf{b} \cdot (\mathbf{c}\mathbf{a}).$$

Recall, the outer product is not commutative!

2. Prove the identities $\mathbf{k} \cdot \overset{a}{\widetilde{\mathbf{P}}} = \mathbf{k} \times \overset{ad}{\widetilde{\mathbf{P}}}$ and $\mathbf{u} \cdot (\mathbf{\nabla}\mathbf{u}) = (\mathbf{u} \cdot \mathbf{\nabla})\mathbf{u}$.

3. From the microscopic definition of the spin angular momentum, Eq. (2.99), derive the dynamics in Eq. (2.100).

4. Use the non-advective form of the momentum balance equation,

$$\frac{\partial \rho \mathbf{u}}{\partial t} = -\mathbf{\nabla} \cdot \mathbf{P},$$ (2.127)

to show that, for this case, we have

$$i\mathbf{k} \cdot \widetilde{\mathbf{P}}(\mathbf{k}, t) = -\sum_i \left(\mathbf{F}_i - i\mathbf{k} \cdot \mathbf{v}_i \mathbf{v}_i \right) e^{-i\mathbf{k} \cdot \mathbf{r}_i}.$$

Give explicit expressions for the shear pressure component $\widetilde{P}_{yx}$ and the normal pressure component $\widetilde{P}_{yy}$ when $\mathbf{k} = (0, k_y, 0)$.

5. For fluid mixtures, we define the mass density of, say, the A component by

$$\rho_A(\mathbf{r}, t) = \sum_{i \in \mathcal{A}} m_i \delta(\mathbf{r} - \mathbf{r}_i).$$

$i \in \mathcal{A}$ symbolises that index i runs over the set $\mathcal{A}$ composed of the indices of A particles. From the $\mathcal{G}$-operator, show the mass balance equation for A is

$$\frac{\partial \rho_A}{\partial t} = -\mathbf{\nabla} \cdot \mathbf{j}_A,$$

where $\mathbf{j}_A = \sum_{i \in \mathcal{A}} m_i \mathbf{v}_i \delta(\mathbf{r} - \mathbf{r}_i)$.

Then derive the balance equation for $\mathbf{j}_A$ using the $\mathcal{H}$-operator and show that $\mathbf{j}_A$ is a non-conserved quantity.

3

Nanoscale Hydrodynamic Relaxations

At first sight, one molecule's thermal motion appears uncorrelated with the motion of other molecules further away as well as the motion at earlier times. This is, in fact, not the case. The molecular motion, and more generally the system dynamic quantities, are indeed correlated with the system's dynamics at other points and earlier times. This phenomenon is referred to as spatio-temporal correlations. For the fluids we study here, the correlations are short ranged, both with respect to time and space, often on the order of picoseconds and nanometres.

Depending on the actual quantity we study, the correlations can be fingerprints of different underlying hydrodynamic processes, and in this chapter we will explore this topic in detail. In doing so we not only gain insight into the hydrodynamic processes behind the correlations, we can also perform a highly controlled exploration of the validity of hydrodynamics on the nanoscale.

The idea of applying hydrodynamics on these small scales is founded in Onsager's regression hypothesis from 1931. This states that the regression of microscopically induced fluctuations in equilibrium follows, on average, the macroscopic laws of small non-equilibrium disturbances [168]. Thus, thermally induced perturbations relax according to the hydrodynamic equations. The relaxations here refer not to quantities like mass density directly, as they are constantly fluctuating and not relaxing, but to the decay of the associated correlation functions [3, 31, 147, 182]. Therefore, we need to derive these hydrodynamic correlation functions and compare the predictions with experimental data (where available) or molecular dynamics results. Interpretation of small-scale correlations on the basis of hydrodynamics is a well-known trade, and Boon and Yip [31] call this *Molecular Hydrodynamics*.

As an introductory example, we will explore how the density for a single-component fluid at a point $\mathbf{r}_1$ and at time t_1 is correlated with the density at point $\mathbf{r}_2$ at time t_2; see the illustration in Fig. 3.1. We define the correlations through the famous van Hove correlation function G [156],

$$\rho_{av} m G(\mathbf{r}_1, t_1, \mathbf{r}_2, t_2) = \langle \rho(\mathbf{r}_1, t_1) \rho(\mathbf{r}_2, t_2) \rangle, \qquad (3.1)$$

where ρ_{av} is the system's average density, and m the molecular mass. Notice that the correlations are given via the expected value of the product $\rho(\mathbf{r}_1, t_1) \rho(\mathbf{r}_2, t_2)$, as indicated with brackets $\langle \ldots \rangle$. In practise we approximate the expected value with an ensemble sample average over a sufficiently large set of independent and equally probable initial conditions. Also, see Appendix A.3 for further details.

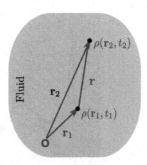

Figure 3.1   Illustration of the densities located at $\mathbf{r}_1$ and $\mathbf{r}_2$, and at two different times, t_1 and t_2.

If the system is homogeneous and isotropic, then the system is both translational and rotational invariant. This means that only the distance between the two points, $r = ||\mathbf{r}_2 - \mathbf{r}_1||$, is relevant. Furthermore, for time-translational invariance, it is the time difference, $t = t_2 - t_1$, which is the relevant variable. Sometimes these invariances are implicitly stated by simply letting, for example, $\mathbf{r}_1 = \mathbf{0}$ and $t_1 = 0$. The invariances means that we can write the van Hove correlation function as

$$\rho_{\text{av}} m G(r,t) = \langle \rho(r,t)\rho(0,0) \rangle. \tag{3.2}$$

Since we are interested in the density fluctuations, the density itself is decomposed into a sum of the average density, ρ_{av}, and the fluctuation part, $\delta\rho$, that is,

$$\rho = \rho_{\text{av}} + \delta\rho. \tag{3.3}$$

The ensemble average is

$$\langle \rho \rangle = \langle \rho_{\text{av}} + \delta\rho \rangle = \langle \rho_{\text{av}} \rangle + \langle \delta\rho \rangle = \rho_{\text{av}} \tag{3.4}$$

due to linearity and since $\langle \delta\rho \rangle = 0$ by definition. We can now interpret the average density ρ_{av} as the ensemble averaged density. In equilibrium this is equivalent with the time average. This is a general decomposition we use for all the hydrodynamic variables; however, we will for purposes of clarity use both the subscript 'av' and the bracket notation $\langle \ldots \rangle$. The van Hove function is now written as

$$\begin{aligned} \rho_{\text{av}} m G(r,t) &= \langle (\rho_{\text{av}} + \delta\rho(r,t))(\rho_{\text{av}} + \delta\rho(0,0)) \rangle \\ &= \rho_{\text{av}}^2 + c_{\rho\rho}(r,t), \end{aligned} \tag{3.5}$$

where the density fluctuation correlation function $c_{\rho\rho}$ is defined as

$$c_{\rho\rho}(r,t) = \langle \delta\rho(r,t)\delta\rho(0,0) \rangle. \tag{3.6}$$

For this introductory example, we simply calculate $c_{\rho\rho}$ for a Lennard–Jones fluid and divide the system into slabs and where the local fluctuating part of the density, $\delta\rho_{\mathcal{V}}$, is then given from the definition Eq. (1.19),

$$\delta\rho_{\mathcal{V}}(t) = \frac{1}{\Delta\mathcal{V}} \sum_{i \in \mathcal{V}} m_i - \rho_{\text{av}}, \tag{3.7}$$

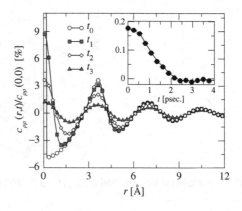

Figure 3.2   Density autocorrelation function as function distance for Lennard–Jones model methane at the liquid state point. Four time snapshots are shown: t_n, $n = 0, 1, 2, 3$, where $t_0 < t_1 < t_2 < t_3$. Insert: Correlation at $r \approx 10\,\text{Å}$ as function of time.

where ΔV is the fluid slab volume and i runs over particles in the slab V. The correlation with respect to r can be evaluated as the correlation between slabs where the slab midpoint represents r. While this calculation fulfils the purpose, it is rather heuristic and we will see a more careful treatment later.

Figure 3.2 shows the density correlation function $c_{\rho\rho}$ as a function of distance r and time t. As expected, the correlation decreases as time increases, and eventually goes to zero as the density at long times becomes completely uncorrelated with the density at $t = 0$. As seen, there exists a sharp peak for $r = 0$ that broadens with time; this happens through a diffusive process and is the so-called self-part of the van Hove function. We return to this in the last part of the chapter.

With respect to spatial coordinate r, the correlation function features damped oscillation with a characteristic length scale of around one molecular diameter, that is, around 3.7 Å. These oscillations were first observed in computer simulations by Rahman in 1964 [182]; they are due to the molecular packing present in the liquid and are referred to as the liquid structure. This structure is mostly pronounced on very small length scales. The insert in Figure 3.2 plots the correlation function at $r \approx 10$ Å, or 1 nm, as a function of time; that is, even on the nanometre length scale correlations exist, which is the focus of our exploration.

As outlined in Chapter 2, rather than investigating the dynamics in real space one often studies the correlations of the corresponding Fourier coefficients. We will write this as

$$\rho_{\text{av}} m F(k,t) = \langle \widetilde{\delta\rho}(k,t)\widetilde{\delta\rho}(-k,0) \rangle , \tag{3.8}$$

where F is the so-called coherent intermediate scattering function (a name that comes from the experimental method), and k is the wavevector magnitude.

Now, to complicate things a bit further, scattering experiments do not provide the scattering function F directly, that is, the time dependence of the Fourier coefficients, but the frequency dependency. This is the dynamic structure factor (even if

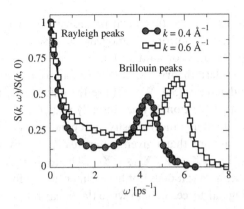

Figure 3.3 Normalised dynamic structure factor of liquid caesium at $T = 308$ K for two different wavevectors. Lines serve as a guide to the eye. Data are taken from Bodensteiner et al. [27]

it is a function), and we write it as $S = S(k, \omega)$, with ω being the angular frequency. Figure 3.3 shows experimental results for the dynamic structure factor for liquid caesium by Bodensteiner et al. [27] Notice that the length scales in the spectra are $2\pi/0.4$ Å^{-1} and $2\pi/0.6$ Å^{-1}, that is, 1.57 and 1.05 nm, respectively.

The dynamic structure factor features two peaks; one peak centred around zero frequency and one peak at non-zero frequency. The former is referred to as the Rayleigh peak and the latter as the Brillouin peak. In this chapter we will explore these types of spectra in detail using the hydrodynamic model and, based on this, make interpretations of the underlying physical processes.

At this point there is an interesting result we can give right away: for fluid systems composed of point mass particles like methane, the classical hydrodynamic theory can predict, at least qualitatively, the correlation functions (and their spectra) down to surprisingly small length and time scales. On extremely small scales, the classic hydrodynamic model is too simple and will be unsatisfactory, and a multiscale model that includes the relevant physical processes must be formulated. These theories are outside the scope of this chapter, as we focus on the application of classical hydrodynamics only.

3.1 Classical Hydrodynamics

The balance equations derived in Chapter 2 do not form closed mathematical problems because the fluxes are unknown functions of the hydrodynamic variables – even if we have a microscopic interpretation of them. However, experimentally we know that the fluxes can be functions of the system gradients. For example, Fourier's law of heat conduction states that the heat flux is proportional to the gradient of the temperature. In the context of the thermal kinetic energy flux, Eq. (2.92), this means that $\mathbf{J}^\varepsilon \propto \nabla(\rho\varepsilon)$. The proportionality is a linear model, or a first-order expansion in the gradient, and we must expect that this linear relationship only holds for sufficiently small gradients. In the case of isotropic homogeneous systems, the coefficient of proportionality is represented by a constant scalar.

The modern theory[1] treating this in a formal manner is irreversible non-equilibrium thermodynamics [52]. We will use the empirically founded approach, where the constants of proportionality are directly related to the (scalar) transport coefficients like heat conductivity and viscosity. Independent of the theoretical framework, we can, in general, assume that a given scalar flux J_i of a given quantity is depending on a set of driving forces $\mathbb{X} = (\mathbb{X}_1, \mathbb{X}_2, \ldots, \mathbb{X}_N)$. The term driving force is here used to indicate that $\mathbb{X}_j$ causes a flux; it does not have dimension of force (or force density) and is therefore not a regular force. We can write the linear relationship in general as

$$J_i = \sum_j M_{ij} \mathbb{X}_j, \tag{3.9}$$

where M_{ij} is the constant of proportionality with units depending on the driving force, and j runs over all N driving forces. Equation (3.9) is easily extended to vector fluxes and so forth. In most nanoscale fluid systems, the linear relationship between the flux and driving forces suffices; however, in some cases, for example, for gas flows, higher-order terms in the gradient expansion are needed, leading to the Burnett equations. We do not consider these situations.

Important note: in this treatment, M_{ij} are not phenomenological coefficients known from thermodynamics and do not follow Onsager's reciprocal relation in general, that is, $M_{ij} \neq M_{ji}$.[2] However, we will still use the fundamental result that if a driving force $\mathbb{X}_j$ leads to flux J_i, we also have that the force $\mathbb{X}_i$ leads to a flux J_j.

One natural next question to ask is what driving forces cause a given flux. According to Curie's principle [52], the forces and the flux must have the same tensorial character; see page 26. Thus, if a flux is a vector, the driving force must also be a vector. Aside from this restriction, a flux can, in principle, be dependent on any driving force present in the system. A few examples:

- In fluid mixtures the fluxes of the different constituents depend not only on the gradients of the concentrations, but also on the temperature gradient [183]. The total mass flux of one component is therefore given as a sum of two gradients: a concentration gradient and a temperature gradient. This is called thermophoresis or the Soret effect.

- For dielectric materials the polarisation rate of change (which we can think of as a flux) depends not only on the electric potential gradient, but also on the temperature gradient [35]; this is relevant for nanoscale systems, where large temperature gradients can be achieved.

- As shown in Chapter 1, for molecular fluids there exists a coupling between the spin angular momentum (a pseudo-vector) and the linear momentum (a vector). We shall

---

[1] Perhaps it is appropriate to reformulate this as "The *most* modern theory."
[2] Not to be confused with Onsager's regression hypothesis.

see later that even though the two quantities have unlike tensorial character, the coupling can still be realised under the constraint of Curie's principle.

The underlying thermal fluctuations of the fluid leaves the hydrodynamic quantities as random variables. To model this, we introduce the stochastic force method [56, 140]. Basically, we add a stochastic term δJ_i to Eq. (3.9); that is, in the linear regime we get for the flux J_i

$$J_i = \sum_j M_{ij} \mathbb{X}_j + \delta J_i. \tag{3.10}$$

The term δJ_i is called the fluctuating dissipative flux [56]; however, we will frequently refer to it as simply the forcing term. For our purpose it suffices that the forcing term has zero mean, and the Fourier coefficients of the force are uncorrelated with the Fourier coefficients of the hydrodynamic variables $\widetilde{A}$, at all times and wave-vectors; that is,

$$\langle \delta J_i \rangle = 0 \quad \text{and} \quad \left\langle \widetilde{\delta J_i}(\mathbf{k},t) \widetilde{A}(-\mathbf{k},t') \right\rangle = 0. \tag{3.11}$$

Again, the brackets denote the average over an ensemble of independent initial conditions.

We study the fluctuations of the hydrodynamic quantities in equilibrium. As we have seen for the density, these quantities are written as a sum of the average and fluctuating parts. In general, we write $A = \rho\phi = (\rho\phi)_{\text{av}} + \delta(\rho\phi)$ or, by writing out the terms,

$$A = (\rho_{\text{av}} + \delta\rho)(\phi_{\text{av}} + \delta\phi)$$
$$\approx \rho_{\text{av}}\phi_{\text{av}} + \rho_{\text{av}}\delta\phi + \phi_{\text{av}}\delta\rho \tag{3.12}$$

to first order in the fluctuations. Comparing terms, we then identify $(\rho\phi)_{\text{av}} = \rho_{\text{av}}\phi_{\text{av}}$ and $\delta(\rho\phi) \approx \rho_{\text{av}}\delta\phi + \phi_{\text{av}}\delta\rho$. The rate of change of A reads

$$\frac{\partial \rho\phi}{\partial t} = \frac{\partial \delta\rho\phi}{\partial t} = \rho_{\text{av}}\frac{\partial \delta\phi}{\partial t} + \phi_{\text{av}}\frac{\partial \delta\rho}{\partial t} \tag{3.13}$$

since the ensemble average is constant in equilibrium (but not in general!). The advective term in the balance equation is of second order with respect to the fluctuations; hence, this can be ignored in the linear regime. The linearised balance equation is in terms of the fluctuations then, in general,

$$\frac{\partial \delta(\rho\phi)}{\partial t} = \sigma - \boldsymbol{\nabla}\cdot\mathbf{J}, \tag{3.14}$$

and if $\phi_{\text{av}} = 0$,

$$\rho_{\text{av}}\frac{\partial \delta\phi}{\partial t} = \sigma - \boldsymbol{\nabla}\cdot\mathbf{J}. \tag{3.15}$$

Since the fluxes are linear with respect to the hydrodynamic variables, we then have only linear problems, which simplifies the mathematical problem significantly.

3.1.1 Dynamical Equations for the Fluctuations

The starting point for our discussion will be fluids composed of point mass molecules. For these systems the antisymmetric part of the pressure tensor is zero and the coupling to the spin angular momentum can be ignored in the next chapter we return to the more general case.

In equilibrium the average streaming velocity is zero, $\mathbf{u}_{av} = \mathbf{0}$, and we have for the linear momentum density

$$\mathbf{j} = (\rho_{av} + \delta\rho)\delta\mathbf{u} \approx \rho_{av}\delta\mathbf{u} \tag{3.16}$$

to first order in the fluctuations. Recall, the mass balance equation

$$\frac{\partial \rho}{\partial t} = -\nabla \cdot \mathbf{j}. \tag{3.17}$$

Substitution of Eq. (3.16) gives

$$\frac{\partial \delta\rho}{\partial t} = -\rho_{av}\nabla \cdot \delta\mathbf{u}. \tag{3.18}$$

Likewise, for the linear momentum density, Eq. (2.62), we get

$$\rho_{av}\frac{\partial \delta\mathbf{u}}{\partial t} = -\nabla \cdot \left((p_{eq} + \Pi)\mathbf{I} + \overset{os}{\mathbf{P}}\right) \tag{3.19}$$

since the advective term is non-linear and we have no external forces acting on the system. Finally, we also need the balance equation for the thermal kinetic energy density fluctuations (or equivalently the kinetic temperature, if you prefer). The average thermal kinetic energy is non-zero and we have that

$$\rho\varepsilon = (\rho\varepsilon)_{av} + \delta(\rho\varepsilon). \tag{3.20}$$

Again, in equilibrium we have that $(\rho\varepsilon)_{av}$ is constant, and substitution into Eq. (2.90) gives

$$\frac{\partial \delta(\rho\varepsilon)}{\partial t} = \frac{T\beta_V}{\rho_{av}}\frac{\partial \delta\rho}{\partial t} - \nabla \cdot \mathbf{J}^{\varepsilon}. \tag{3.21}$$

Equations (3.18), (3.19), and (3.21) are the linearised balance equations for the fluctuations.

To proceed, we assume homogeneity and isotropy. The fluxes, Π, $\overset{os}{\mathbf{P}}$, and $\mathbf{J}^{\varepsilon}$, are modelled through classic hydrodynamic constitutive relations, but with the addition of stochastic forcing; the general form is given in Eq. (3.10):

$$\Pi = -\eta_v(\nabla \cdot \mathbf{u}) + \delta\Pi \tag{3.22a}$$

$$\overset{os}{\mathbf{P}} = -2\eta_0(\overset{os}{\nabla\mathbf{u}}) + \delta\overset{os}{\mathbf{P}} = -\eta_0\left((\nabla\mathbf{u} + \mathbf{u}\nabla) - \frac{2}{3}\mathrm{Tr}(\nabla \cdot \mathbf{u})\mathbf{I}\right) + \delta\overset{os}{\mathbf{P}} \tag{3.22b}$$

$$\mathbf{J}^{\varepsilon} = -\frac{\lambda}{c_V\rho_{av}}\nabla(\rho\varepsilon) + \delta\mathbf{J}^{\varepsilon}, \tag{3.22c}$$

where η_v and η_0 are the bulk and shear viscosities, and λ is the heat conductivity. Notice that Curie's constraint is obeyed and that the constitutive models are simply Newton's law of viscosity and Fourier's law of heat conduction with added stochastic forcing.

The reader may be more familiar with Newton's law written in terms of the strain rate, $\overset{os}{\dot{\gamma}} = \boldsymbol{\nabla}\mathbf{u}$. For completeness we list this formulation as well; thus, for isotropic systems the symmetric pressure tensor reads

$$\overset{s}{\mathbf{P}} = p_{eq}\mathbf{I} - \eta_v(\boldsymbol{\nabla}\cdot\mathbf{u})\mathbf{I} - 2\eta_0\overset{s}{\dot{\gamma}}. \tag{3.23}$$

Moving on, using the identities from vector calculus,

$$\boldsymbol{\nabla}\cdot(\overset{os}{\boldsymbol{\nabla}\mathbf{u}}) = \frac{1}{2}\nabla^2\mathbf{u} + \frac{1}{2}\boldsymbol{\nabla}(\boldsymbol{\nabla}\cdot\mathbf{u}) \tag{3.24a}$$

$$\boldsymbol{\nabla}\cdot a\mathbf{I} = \boldsymbol{\nabla}a, \tag{3.24b}$$

where a is a real scalar, we have the linearised dynamical equations for the fluctuations

$$\frac{\partial\delta\rho}{\partial t} = -\rho_{av}\boldsymbol{\nabla}\cdot\delta\mathbf{u} \tag{3.25a}$$

$$\rho_{av}\frac{\partial\delta\mathbf{u}}{\partial t} = -\boldsymbol{\nabla}\delta p_{eq} + (\eta_v + \eta_0/3)\boldsymbol{\nabla}(\boldsymbol{\nabla}\cdot\delta\mathbf{u}) + \eta_0\nabla^2\delta\mathbf{u} - \boldsymbol{\nabla}\cdot\delta\mathbf{P} \tag{3.25b}$$

$$\frac{\partial\delta(\rho\varepsilon)}{\partial t} = \frac{T\beta_V}{\rho_{av}}\frac{\partial\delta\rho}{\partial t} + \frac{\lambda}{c_V\rho_{av}}\nabla^2\delta(\rho\varepsilon) - \boldsymbol{\nabla}\cdot\delta\mathbf{J}^\varepsilon, \tag{3.25c}$$

where $\delta\mathbf{P} = (\delta\Pi)\mathbf{I} + \delta\overset{os}{\mathbf{P}}$.

In our treatment of equilibrium relaxations we are interested in the Fourier coefficients of the dynamical variables. By Fourier transforming, see Eq. (2.6a), we get

$$\frac{\partial\widetilde{\delta\rho}}{\partial t} = -i\rho_{av}\mathbf{k}\cdot\widetilde{\delta\mathbf{u}} \tag{3.26a}$$

$$\rho_{av}\frac{\partial\widetilde{\delta\mathbf{u}}}{\partial t} = -i\mathbf{k}\cdot\widetilde{\delta p}_{eq} - (\eta_v + \eta_0/3)\mathbf{k}(\mathbf{k}\cdot\widetilde{\delta\mathbf{u}}) - \eta_0 k^2\widetilde{\delta\mathbf{u}} - i\mathbf{k}\cdot\widetilde{\delta\mathbf{P}} \tag{3.26b}$$

$$\frac{\partial\widetilde{\delta(\rho\varepsilon)}}{\partial t} = \frac{T\beta_V}{\rho_{av}}\frac{\partial\widetilde{\delta\rho}}{\partial t} - \frac{\lambda}{\rho_{av}c_V}k^2\widetilde{\delta(\rho\varepsilon)} - i\mathbf{k}\cdot\widetilde{\delta\mathbf{J}}^\varepsilon. \tag{3.26c}$$

Thus, these equations for the density, momentum, and energy Fourier coefficients form our classical hydrodynamic model of the fluid system at equilibrium.

3.1.2 Transverse Relaxations

One can make a simple choice for the wavevector (or equivalently the system coordinate) such that the dynamics will decompose into one parallel mode (longitudinal dynamics) and two perpendicular modes (transverse dynamics). For example, let $\mathbf{k} = (0, k_y, 0)$; then the streaming velocity x and z components, denoted the transverse components, read from Eq. (3.26b) as follows:

$$\rho_{av}\frac{\partial \widetilde{\delta u_x}}{\partial t} = -\eta_0 k_y^2 \widetilde{\delta u_x} - ik_y \widetilde{\delta P_{yx}} \qquad (3.27a)$$

$$\rho_{av}\frac{\partial \widetilde{\delta u_z}}{\partial t} = -\eta_0 k_y^2 \widetilde{\delta u_z} - ik_y \widetilde{\delta P_{yz}}. \qquad (3.27b)$$

Importantly, these Fourier coefficients decouple from the dynamics of the other hydrodynamic variables (i.e., density and kinetic energy) and therefore pose a particular simple problem. Eqs. (3.27a) and (3.27b) are identical, and the dynamics depend on the same fluid property, namely, the shear viscosity. Thus, the transverse dynamics is a purely shear viscous dynamics. We can simply continue by using one of the two transverse dynamical equations, say Eq. (3.27a), to study the transverse dynamics.

As we have noted, we do not study the hydrodynamics directly via Eq. (3.27a), but through the corresponding hydrodynamic correlation function, namely, the transverse velocity autocorrelation function, $C_{uu}^{\perp}$:

$$C_{uu}^{\perp}(\mathbf{k},t) = \frac{1}{V}\left\langle \widetilde{\delta u_x}(\mathbf{k},t)\widetilde{\delta u_x}(-\mathbf{k},0)\right\rangle. \qquad (3.28)$$

In a more rigorous treatment [31], one will study the transverse linear momentum autocorrelation; however, to first order in the fluctuations Eq. (3.28) suffices.

To derive the equation for $C_{uu}^{\perp}$, Eq. (3.27a) is multiplied by $\widetilde{\delta u_x}(-k_y,0)$, and we perform an ensemble average over independent initial conditions. First, notice that the left-hand side reads

$$\left\langle \left(\rho_{av}\frac{\partial}{\partial t}\widetilde{\delta u_x}(\mathbf{k},t)\right)\widetilde{\delta u_x}(-\mathbf{k},0)\right\rangle = \rho_{av}\frac{\partial}{\partial t}\left\langle \widetilde{\delta u_x}(\mathbf{k},t)\widetilde{\delta u_x}(-\mathbf{k},0)\right\rangle \qquad (3.29)$$

since $\delta u_x(\mathbf{k},0)$ is constant with respect to time. Also, recall that the stochastic force has the property that

$$\langle \widetilde{\delta P_{yx}}(\mathbf{k},t)\widetilde{\delta u_x}(-\mathbf{k},0)\rangle = 0 \qquad (3.30)$$

and, hence, we arrive at

$$\rho_{av}\frac{\partial}{\partial t}C_{uu}^{\perp}(\mathbf{k},t) = -\eta_0 k_y^2 C_{uu}^{\perp}(\mathbf{k},t). \qquad (3.31)$$

The dynamics for the transverse velocity autocorrelation function is then given by a simple deterministic first-order differential equation. The solution is

$$C_{uu}^{\perp}(\mathbf{k},t) = C_{uu}^{\perp}(\mathbf{k},0)\,e^{-\eta_0 k_y^2 t/\rho_{av}}. \qquad (3.32)$$

The initial condition (or prefactor), $C_{uu}^{\perp}(\mathbf{k},0)$, can be expressed in terms of density and temperature. To see this we write the fluctuations in streaming velocity through the molecular definition. Again, to first order,

$$\rho_{av}\widetilde{\delta u} \approx m\sum_i \mathbf{v}_i\,e^{-i\mathbf{k}\cdot\mathbf{r}_i} \qquad (3.33)$$

for a single-component fluid with the molecular mass m. At $t = 0$,

$$
VC_{uu}^{\perp}(\mathbf{k}, 0) = \left\langle \widetilde{\delta u_x}(\mathbf{k}, 0) \widetilde{\delta u_x}(-\mathbf{k}, 0) \right\rangle
$$

$$
= \frac{m^2}{\rho_{\mathrm{av}}^2} \left\langle \left[\sum_i v_{i,x}(0) e^{-ik_y y_i} \right] \left[\sum_i v_{i,x}(0) e^{ik_y y_i} \right] \right\rangle
$$

$$
= \frac{m^2}{\rho_{\mathrm{av}}^2} \left\langle \sum_i v_{i,x}^2(0) \right\rangle \tag{3.34}
$$

since the microscopic velocities are initially uncorrelated, and the instantaneous cross-correlations are zero. From the equipartition theorem, $\langle \sum_i m v_{i,x}^2(0) \rangle = N k_B T$, where N is the number of particles, and since Nm is total mass we obtain

$$
C_{uu}^{\perp}(\mathbf{k}, 0) = \frac{k_B T}{\rho_{\mathrm{av}}} . \tag{3.35}
$$

We can then write the solution as

$$
C_{uu}^{\perp}(\mathbf{k}, t) = \frac{k_B T}{\rho_{\mathrm{av}}} e^{-\omega_0 t} , \tag{3.36}
$$

where the characteristic transverse frequency (which is the eigenvalue of the problem) is defined from Eq. (3.32) to be

$$
\omega_0 = \frac{\eta_0 k_y^2}{\rho_{\mathrm{av}}} . \tag{3.37}
$$

The fact that the characteristic frequency depends on the wavevector squared is a fingerprint of a diffusion process; in this case, diffusion of linear momentum transverse to the wavevector direction.

We can make a detailed comparison of the theoretical predictions with molecular dynamics simulations, where the correlation function can be calculated directly from the definitions in Eqs. (3.33) and (3.28). In Fig. 3.4 results are plotted for model liquid methane, where the molecular interactions are given through the standard Lennard–Jones potential, Eq. (1.22).

To compare the simulation data with the hydrodynamic predictions without performing any fitting, the methane model's shear viscosity can be evaluated from independent simulations using the Green–Kubo integral [85, 137],

$$
\eta_0 = \frac{V}{k_B T} \int_0^{\infty} \langle P_{0,xy}(t) P_{0,xy}(0) \rangle \, \mathrm{d}t , \tag{3.38}
$$

where $P_{0,xy}(t)$ is the xy-tensor component of the zero wavevector pressure tensor, Eq. (2.52). See Daivis and Evans [47] for an elegant method to use all shear pressure components to calculate the shear viscosity. This integration gives $\eta_0 = 0.15$ mPa·s [185] at the state point chosen. The hydrodynamic predictions are shown in Fig. 3.4 as punctured lines for the two smallest wavevectors. It is observed that for wavevectors $k_y \leq 0.12$ Å^{-1} the hydrodynamic theory is in fair agreement with simulation data, at least for sufficiently long times. This corresponds to wavelengths larger than $2\pi/k_y = 5.2$ nm.

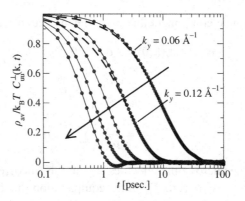

Figure 3.4 Transverse velocity autocorrelation function for model methane liquid at state point $(\rho, T) = (446\,\mathrm{kg m}^{-3}, 166\,\mathrm{K})$. Filled circles connected with lines are simulation data, and the punctured lines are graphs for Eq. (3.36) using $\eta_0 = 0.15\,\mathrm{mPa \cdot s}$. The arrow indicates increasing wavevector; the interval is $k_y = 0.06\,\text{Å}^{-1}$ to $k_y = 0.31$ Å^{-1}. Data are re-plotted from Ref. [91].

The agreement should be quantified in some manner. For example, we here use an error estimator, err, as the mean squared deviation normalised with respect to the squared prediction [91]. If $C_{uu}^{\perp}$ is the true (here from simulation) transverse velocity autocorrelation function and C_{hyd} the hydrodynamic predictions, err is given as

$$\mathrm{err}(k_y) = \frac{\int \left[C_{uu}^{\perp}(k_y, t) - C_{\mathrm{hyd}}(k_y, t) \right]^2 \mathrm{d}t}{\int C_{\mathrm{hyd}}^2(k_y, t)\,\mathrm{d}t}. \tag{3.39}$$

The error estimator is plotted in Fig. 3.5 as a function of wavevector; at $k_y = 0.12\,\text{Å}^{-1}$ we have err = 0.4 per cent. There is no general criterion for a threshold acceptance value or even a standard for the error estimator; hence, determining a hydrodynamic limit is ambiguous.

On a very short time scale, the molecules are bounded by cages formed by the surrounding fluid, and the molecular motion is a rattling one, like atoms in a crystal. This

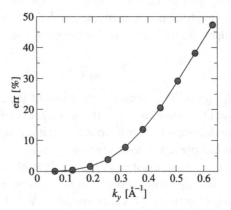

Figure 3.5 Error estimator for the liquid methane model. Data re-plotted from Ref. [91].

leads to solid-like elastic behaviour and allows for additional propagating transverse shear waves [75, 203]. The shear waves are observed as anti-correlations at small length and time scales, see Fig 3.4 for $k_y = 0.31^{-1}$, and thus not predicted by the classical theory, Eq. (3.36), which includes only the presence of a diffusive process.

From this picture we can define a characteristic escape time, τ_F, which is the average time it takes the molecules to escape the fluidic cage [75, 203]; we will denote τ_F the Frenkel time after Y. Frenkel, who made significant contributions to the theory of condensed matter. For timescales larger than τ_F the molecules enter the diffusive regime and the system's dynamics becomes fluidic and viscous. The cage and rattling motion is a rare event for gasses, and we here only discuss the Frenkel time in connection with dense systems, which is also the scope of the book. At the end of the chapter we will return to how τ_F can be estimated.

It is evident from both the theory and simulation data, Fig. 3.4, that the correlation decay time depends on the specific wavevector (i.e., the characteristic length scale we study). At small length scales the relaxation is extremely fast, and the system never enters the fluidic regime. From Eq. (3.37) we define a wavevector-dependent relaxation time, $\tau(k_y) = 2\pi\rho_{av}/(\eta_0 k_y^2)$. If τ is larger than the time it takes for the fluid internal stress to relax, denoted τ_s, then Bocquet and Charlaix [25] conjectured that the continuum approximation is applicable, that is, when

$$\tau > \tau_s \text{ implying } k_y < \sqrt{\frac{2\pi\rho_{av}}{\eta_0 \tau_s}}. \tag{3.40}$$

We refer to this as the Bocquet–Charlaix criterion. Hansen et al. [91, 102] used the time where the pressure tensor autocorrelation, defined in Eq. (3.38), is fully relaxed as an estimator for τ_s. For the liquid methane model, this gives $\tau_s \approx 4$ psec., and we have from the Bocquet–Charlaix criterion that $k_y < 0.27\text{Å}^{-1}$ or a wave-length of approximately 2.3 nm. This corresponds to an error, Eq. (3.39), of around 4 per cent.

Note that the Bocquet–Charlaix criterion pertains to the collective dynamics, whereas the Frenkel escape time is determined by the single-particle escape time.

It is often informative to evaluate the corresponding mechanical spectrum; that is, we transform the correlation function from the time domain into the frequency domain. The transformation is carried out using the Fourier–Laplace transform defined as

$$\mathcal{L}[f(t)] = \widehat{f}(\omega) = \int_0^\infty f(t)\, e^{-i\omega t}\, dt, \tag{3.41}$$

where ω is the angular frequency. The theoretical predicted spectrum for the transverse velocity autocorrelation function is then

$$\widehat{C}_{uu}^\perp(\mathbf{k}, \omega) = \frac{k_B T}{\rho_{av}} \int_0^\infty e^{-(\omega_0 + i\omega)t}\, dt = \frac{k_B T}{\rho_{av}} \frac{1}{\omega_0 + i\omega}. \tag{3.42}$$

The mechanical spectrum has a real part and an imaginary part. The imaginary part of the spectrum for $\widehat{C}_{uu}^\perp$ is

$$\text{Im}\left[\widehat{C}_{uu}^\perp\right] = \frac{k_B T \omega}{\rho_{av}(\omega^2 + \omega_0^2)}, \tag{3.43}$$

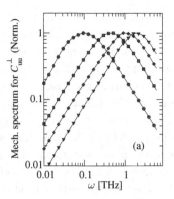
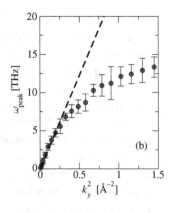

Figure 3.6 (a) Imaginary part of the mechanical spectrum of the transverse velocity autocorrelation function for model liquid methane. The wavevector interval is from $k_y = 0.06$ Å^{-1} to $k_y = 0.25$ Å^{-1}. (b) The dispersion relation curve. Punctured line is the theoretical prediction.

which has maxima (or peaks) at the frequencies $\pm\omega_{\text{peak}}$ with

$$\omega_{\text{peak}} = \omega_0 = \frac{\eta_0 k_y^2}{\rho_{\text{av}}}. \tag{3.44}$$

This relation between the peak frequency and the wavevector is the dispersion relation. Notice that the fingerprint of the diffusive process, namely, the k_y^2-dependency is also found in the dispersion relation.

In Fig. 3.6(a) the imaginary part of the spectrum of $C_{uu}^{\perp}$ is plotted for different wavevectors for the methane model; we only plot the results for positive ω due to the spectrum symmetry. The peak is clearly visible and shifts to higher frequencies with increasing wavevector as expected from the derived dispersion relation, Eq. (3.44). In fact, in the low-wavevector regime it can be seen from Fig. 3.6 (b) that the peak frequency is, within statistical errors, proportional to k_y^2 (punctured line). Using the viscosity found from the Green–Kubo integral, the predicted dispersion relation is also plotted; the slope is given by η_0/ρ_{av}. It is possible to calculate the shear viscosity from the plot; however, this can be associated with relatively large statistical uncertainties, partly due to the numerical Fourier–Laplace transform, and must be done with some care.

One can also explore the real part of the spectrum. However, the real part has the exact same information as the imaginary part and, in fact, the two have an integral relation formalised elegantly by the Kramer–Krönig relation. Nevertheless, sometimes it can be helpful to study both parts, as they highlight different features of the spectrum; for the transverse dynamics, the imaginary part suffices for our purpose.

Moving on to structured molecules, the transverse dynamics for water at ambient conditions has also been studied; see, for example, [19, 91]. For water at ambient conditions, τ_s does not exceed 10 psec., and then, according to the Bocquet–Charlaix criterion, the classical hydrodynamic theory is applicable for $k_y < 0.07$Å^{-1}. This corresponds to a characteristic length scale of 9 nm. Figure 3.7 plots the transverse

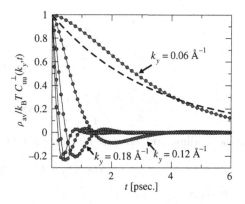

Figure 3.7   Transverse velocity autocorrelation function for ambient water for wavevectors in the interval $k_y = [0.06; 0.30]$ Å^{-1}; the three lowest values are indicated. Symbols connected with lines are molecular dynamics results, and punctured line predictions are from hydrodynamics. Data re-plotted from Ref. [91].

velocity autocorrelation function for water. The smallest wavevector, $k_y = 0.06\text{Å}^{-1}$, fulfils the Bocquet–Charlaix criterion; the error estimator here is again 4 per cent. Note that for large wavevectors the viscoelastic behaviour observed for the point mass Lennard–Jones type systems also is present.

In summary: for simple fluids, classical hydrodynamics successfully describes the transverse relaxations down to the nanoscale and for times above just a few pico-seconds. This result, surprising to many, is due to the large degree of molecular interactions or, equivalently, the frequent momentum exchange events [11]. If σ_{path} is a characteristic mean 'interaction-free' length scale and τ_{path} the timescale between momentum exchange, the hydrodynamic regime is often expressed by [11, 90],

$$\tau_{\text{path}}\omega \ll 1 \quad \text{and} \quad \sigma_{\text{path}}k_y \ll 1 . \tag{3.45}$$

For relatively dense fluids, τ_{path} and σ_{path} are small compared to relatively dilute fluids; hence, the hydrodynamic frequency and wavevector limit are larger for dense fluids. This is in accordance with our discussion in Chapter 1. By cooling the system into the glass state, we can envision that the momentum exchange frequency decreases, increasing τ_{path}; hence the hydrodynamic frequency limit decreases. This is in agree-ment with the Frenkel time picture, where τ_F can be on the order of years. Also, the stress relaxation time τ_s and the viscosity η_0 are many orders of magnitude larger at the glass state compared to the liquid state; hence, according to the Bocquet–Charlaix criterion, both the frequency and the wavevector limit decrease dramatically, and the classical hydrodynamic theory will be grossly unsatisfactory at the nanoscale.

3.1.3 Longitudinal Relaxations

Recall Eq. (3.26). If we again choose the wavevector $\mathbf{k} = (0, k_y, 0)$, we obtain a set of coupled differential equations for the density, y-component streaming velocity, and thermal kinetic energy, namely,

$$\frac{\partial \widetilde{\delta\rho}}{\partial t} = -i\rho_{av}k_y\widetilde{\delta u_y} \tag{3.46a}$$

$$\frac{\partial \widetilde{\delta u_y}}{\partial t} = -\frac{ik_y}{\rho_{av}}\widetilde{\delta p_{eq}} - v_l k_y^2 \widetilde{\delta u_y} - \frac{ik_y}{\rho_{av}}\widetilde{\delta P_{yy}} \tag{3.46b}$$

$$\frac{\partial \widetilde{\delta(\rho\varepsilon)}}{\partial t} = \frac{T\beta_V}{\rho_{av}}\frac{\partial \widetilde{\delta\rho}}{\partial t} - \kappa k_y^2 \widetilde{\delta(\rho\varepsilon)} - ik_y\widetilde{\delta J_y^\varepsilon}. \tag{3.46c}$$

To ease the reading a bit, we have introduced the longitudinal viscosity,

$$v_l = (\eta_v + 4\eta_0/3)/\rho_{av} \tag{3.47}$$

and $\kappa = \lambda/(c_V\rho_{av})$.

To proceed, we (again) assume local thermodynamic equilibrium such that the pressure fluctuations appearing in Eq. (3.46b) can be written as a function of the density and kinetic temperature fluctuations; that is, $\delta p_{eq} = \delta p_{eq}(\delta\rho, \delta T)$. For small fluctuation amplitudes, we expand to first order,

$$\widetilde{\delta p_{eq}} = \left(\frac{\partial p}{\partial \rho}\right)_T \widetilde{\delta\rho} + \left(\frac{\partial p}{\partial T}\right)_\rho \widetilde{\delta T}. \tag{3.48}$$

Recall, the subscripts indicate that the temperature and density are fixed during the differentiation. If the number of particles is fixed, then by the chain rule

$$\left(\frac{\partial p}{\partial \rho}\right)_T = -\frac{V}{\rho_{av}}\left(\frac{\partial p}{\partial V}\right)_T = \frac{1}{\rho_{av}\beta_T}, \tag{3.49}$$

where $\beta_T = -(\partial V/\partial p)_T/V$ is the isothermal compressibility. Furthermore, the thermal pressure coefficient is defined as $\beta_V = (\partial p/\partial T)_\rho$ and we obtain for the pressure fluctuations (now in terms of the thermal kinetic energy)

$$\widetilde{\delta p_{eq}} = \frac{1}{\rho_{av}\beta_T}\widetilde{\delta\rho} + \frac{\beta_V}{\rho_{av}c_V}\widetilde{\delta(\rho\varepsilon)}. \tag{3.50}$$

The thermal kinetic energy depends on the density, see Eq. (3.46c), through its time derivative. This is not the convenient form, however; from the mass balance equation we have

$$\frac{T\beta_V}{\rho_{av}}\frac{\partial \widetilde{\delta\rho}}{\partial t} = -ik_y T\beta_V\widetilde{\delta u_y}; \tag{3.51}$$

thus, substituting Eqs. (3.50) and (3.51) into Eq. (3.46), we arrive at

$$\frac{\partial \widetilde{\delta\rho}}{\partial t} = -i\rho_{av}k_y\widetilde{\delta u_y} \tag{3.52a}$$

$$\frac{\partial \widetilde{\delta u_y}}{\partial t} = -\frac{ik_y}{\rho_{av}^2\beta_T}\widetilde{\delta\rho} - v_l k_y^2 \widetilde{\delta u_y} - \frac{ik_y\beta_V}{c_V\rho_{av}^2}\widetilde{\delta(\rho\varepsilon)} - \frac{ik_y}{\rho_{av}}\widetilde{\delta P_{yy}} \tag{3.52b}$$

$$\frac{\partial \widetilde{\delta(\rho\varepsilon)}}{\partial t} = -iT\beta_V k_y\widetilde{\delta u_y} - \kappa k_y^2 \widetilde{\delta(\rho\varepsilon)} - ik_y\widetilde{\delta J_y^\varepsilon}. \tag{3.52c}$$

These equations are the dynamical equations for the longitudinal Fourier coefficients.

From Eqs. (3.52) we can form nine correlation functions, three from each of the Eqs. (3.52a)–(3.52c). For example, dynamical equations for the density–density, $C_{\rho\rho}$, the density–velocity, $C_{\rho u}$, and the density–energy, $C_{\rho e}$, correlation functions are defined from multiplying Eq. (3.52a) by $\widetilde{\delta\rho}(-k_y,0)$, $\widetilde{\delta u_y}(-k_y,0)$, and $\widetilde{\delta(\rho\varepsilon)}(-k_y,0)$, respectively, and performing the usual ensemble average over independent initial conditions. The dynamical equations for the correlation functions can be written in a compact matrix notation as

$$\frac{\partial}{\partial t}\begin{bmatrix} C_{\rho\rho} & C_{\rho u} & C_{\rho e} \\ C_{u\rho} & C_{uu} & C_{ue} \\ C_{e\rho} & C_{eu} & C_{ee} \end{bmatrix} = -\begin{bmatrix} 0 & i\rho_{av}k_y & 0 \\ \frac{ik_y}{\rho_{av}^2\chi_T} & \nu_l k_y^2 & \frac{ik_y\beta_V}{c_V\rho_{av}} \\ 0 & \frac{iT\beta_V k}{\rho_{av}} & \kappa k_y^2 \end{bmatrix} \cdot \begin{bmatrix} C_{\rho\rho} & C_{\rho u} & C_{\rho e} \\ C_{u\rho} & C_{uu} & C_{ue} \\ C_{e\rho} & C_{eu} & C_{ee} \end{bmatrix}. \tag{3.53}$$

The coefficient matrix is referred to as the hydrodynamic matrix. From Eq. (3.53) we can identify three sets of co-dependent functions, namely,

$$\{C_{\rho\rho}, C_{u\rho}, C_{e\rho}\}, \{C_{uu}, C_{\rho u}, C_{eu}\}, \text{ and } \{C_{ee}, C_{ue}, C_{\rho e}\}. \tag{3.54}$$

Each set then forms a closed three-dimensional problem wherein the constant coefficient matrix is given by the hydrodynamic matrix, that is, the eigenvalues of the problem are the eigenvalues of the hydrodynamic matrix. The general solution to such a linear three-dimensional problem is a sum of three exponential functions; that is, for each correlation function the solution can be written as

$$C(k_y,t) = C_1 e^{\omega_1(k_y)t} + C_2 e^{\omega_2(k_y)t} + +C_3 e^{\omega_3(k_y)t}, \tag{3.55}$$

where C_1, C_2, and C_3 are determined by the initial conditions.

It can be verified that the discriminant of the characteristic polynomial associated with the hydrodynamic matrix is positive. This means the hydrodynamic matrix always has one real and two complex conjugated eigenvalues. Without going through too much tedious algebra, we quickly list the results for these eigenvalues up to second order in wavevector

$$\omega_1 = D_T k_y^2 + \ldots \tag{3.56a}$$

$$\omega_{2,3} = \pm i c_s k_y + \Gamma k_y^2 + \ldots. \tag{3.56b}$$

Here c_s is the adiabatic speed of sound given by

$$c_s^2 = \frac{\beta_V^2\beta_T T + \rho_{av}c_V}{\beta_T c_V \rho_{av}^2}, \tag{3.57}$$

D_T is the thermal diffusivity

$$D_T = \frac{\lambda}{\beta_V^2\beta_T T + \rho_{av}c_V}, \tag{3.58}$$

and Γ the sound attenuation coefficient

$$\Gamma = \frac{1}{2}[\nu_l + \kappa - D_T]. \tag{3.59}$$

The coefficients can be written in a more attractive (and indeed standard) form by introducing the ratio of heat capacities,

$$\gamma = \frac{c_p}{c_V} = 1 + \frac{T\beta_T\beta_V^2}{\rho_{\mathrm{av}}c_V}, \tag{3.60}$$

where c_p is the specific heat capacity at constant normal pressure. This gives

$$c_s^2 = \frac{\gamma}{\rho_{\mathrm{av}}\beta_T}, \ D_T = \frac{\lambda}{\rho_{\mathrm{av}}c_p}, \ \text{and } \Gamma = \frac{1}{2}\left(\frac{\gamma-1}{\gamma}\kappa + \nu_l\right). \tag{3.61}$$

For a ratio of specific heats close to unity (which is the case for many real liquids), the system features little thermal expansivity. In this limit the attenuation coefficient is simplified to $\Gamma \approx \nu_l/2$.

Invoking Euler's identity, we can re-write to second order in wavevector the general form, Eq. (3.55), for the nine correlation functions

$$C(k_y,t) = K_1 e^{-D_T k_y^2 t} + e^{-\Gamma k_y^2 t}\left[K_2\cos(c_s k_y t) + iK_3\sin(c_s k_y t)\right], \tag{3.62}$$

where $K_1 = C_1$, $K_2 = C_2 + C_3$, and $K_3 = C_2 - C_3$. We only need to explore a single correlation function, since the hydrodynamic processes are all featured in each function. We here choose the density autocorrelation function, as this is related to the dynamic structure factor, which can be found experimentally. Our construction of the autocorrelation function implies that it is an even function, see Appendix A.3, and the last term in Eq. (3.62) must be zero, $K_3 = 0$. The normalised density autocorrelation function then comes in the form

$$C_{\rho\rho}^N(k_y,t) = \frac{C_{\rho\rho}(k_y,t)}{C_{\rho\rho}(k_y,0)} = \frac{1}{\gamma}\left[(\gamma-1)e^{-D_T k_y^2 t} + e^{-\Gamma k_y^2 t}\cos(c_s k_y t)\right]. \tag{3.63}$$

We see that the classical hydrodynamic theory predicts that:

1. There exists an exponential relaxation process determined by the thermal diffusivity D_T; the process is a diffusive process (seen by the k_y^2 dependency). For a ratio of specific heats of unity, $\gamma = 1$, this process is suppressed. This is the Rayleigh process.
2. There exists an oscillatory relaxation process with a wavevector-dependent frequency, $c_s k_y$. The dampening of the oscillations is given through a diffusive process depending on the attenuation coefficient Γ, that is, the longitudinal viscosity and heat conductivity. The process accounts for the density waves induced by thermal fluctuations and propagates adiabatically (without exchange of heat with the surroundings) through the system. The propagation speed is given by c_s. This is the Brillouin process.

To make a comparison with experimental results, we study the corresponding mechanical spectrum. To this end it is noted that the Fourier–Laplace transformation is linear, that

$$\int_0^\infty e^{-(i\omega + D_T k_y^2)t}\,\mathrm{d}t = \frac{1}{D_T k_y^2 + i\omega}, \tag{3.64}$$

and, that for $\omega \geq 0$,

$$\int_0^\infty e^{-(i\omega + \Gamma k_y^2)t} \cos(c_s kt)\, dt = \frac{i\omega + \Gamma k_y^2}{(i\omega + \Gamma k_y^2)^2 + (c_s k_y)^2}, \tag{3.65}$$

such that we end with the normalised frequency-dependent density correlation function

$$\widehat{C}_{\rho\rho}^N(k_y, \omega) = \frac{1}{\gamma}\left[\frac{\gamma - 1}{D_T k_y^2 + i\omega} + \frac{i\omega + \Gamma k_y^2}{(i\omega + \Gamma k_y^2)^2 + (c_s k_y)^2}\right]. \tag{3.66}$$

The real part of the spectrum, Eq. (3.66), is proportional to the dynamical structure factor S, and from Eq. (3.66) we have

$$S(k_y, \omega) = \frac{\Delta S}{\gamma}\left[\frac{(\gamma - 1)D_T k_y^2}{D_T^2 k_y^4 + \omega^2} + \frac{(c_s^2 k_y^2 + \Gamma^2 k_y^4 + \omega^2)\Gamma k_y^2}{(c_s^2 k_y^2 + \Gamma^2 k_y^4 - \omega^2)^2 + 4\omega^2 \Gamma^2 k_y^4}\right], \tag{3.67}$$

where ΔS is the amplitude. The first term on the right-hand side is the Rayleigh process, and the second term is the Brillouin process. This form is in agreement with the experimental data by Bodensteiner et al., Fig. 3.3.

As for the transverse dynamics, it is possible to make a more detailed comparison with simulation data. To this end, recall that the density Fourier coefficient for general wavevector $\mathbf{k}$, defined microscopically by

$$\widetilde{\rho}(\mathbf{k}, t) = \sum_i m_i e^{-\mathbf{k}\cdot\mathbf{r}_i}, \tag{3.68}$$

and the density correlation function may then be calculated directly from this definition. Figure 3.8 (a) plots molecular dynamics simulation data for the normalised density autocorrelation function $C_{\rho\rho}^N$ at three different wavevectors; the system is the usual model methane liquid. It is observed that the correlation function feature's damped oscillations superimposed an exponential decay, as predicted by the theory. Both the exponential decay rate and the dampening of the oscillations we expect from the theory to increase with increasing wavevector, and again this also is observed in the simulations.

Figure 3.8(b) shows the corresponding spectrum, and the Rayleigh and Brillouin processes are immediately clear. The spectrum is remarkably similar (qualitatively) to the experimental findings (see Fig. 3.3), and we now understand the underlying hydrodynamic processes behind the spectrum features.

A direct and quantitative comparison between the theory and data is not straightforward as it was for the transverse dynamics because of the very large number of parameters involved in the longitudinal dynamics. However, we can compare predicted dispersion relations with molecular dynamics results. We choose two such relations: (i) The Rayleigh half-peak width and (ii) the Brillouin peak frequency. For the former we first explicitly write the Rayleigh process term, that is, the first term on the right-hand side of Eq. (3.67), as

$$R(\mathbf{k}, \omega) = \frac{\Delta S}{\gamma}\frac{(\gamma - 1)D_T k_y^2}{D_T^2 k_y^4 + \omega^2}. \tag{3.69}$$

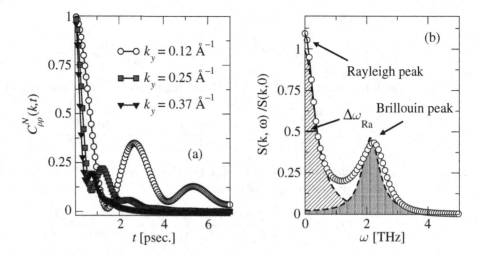

Figure 3.8 Molecular dynamics simulations of liquid methane at state point $(\rho, T) = (446\,\mathrm{kgm^{-3}}; 166\,\mathrm{K})$. (a) The normalised density autocorrelation function $C_{\rho\rho}^{N}$ at different wavevectors. Lines serve as a guide to the eye. (b) The corresponding dynamical structure factor for wavevector $k_y = 0.12\,\text{Å}^{-1}$. The shaded regions under the punctured lines represent the Rayleigh and Brillouin processes found from fitting Eq. (3.63) to data and then using the parameters in Eq. (3.67).

From this we can define the half-peak width $\Delta\omega_{\mathrm{Ra}}$ as

$$\frac{R(\mathbf{k}, \Delta\omega_{\mathrm{Ra}})}{R(\mathbf{k}, 0)} = \frac{D_T^2 k_y^4}{D_T^2 k_y^4 + \Delta\omega_{\mathrm{Ra}}^2} = \frac{1}{2}, \tag{3.70}$$

implying

$$\Delta\omega_{\mathrm{Ra}} = D_T k_y^2. \tag{3.71}$$

This is the first dispersion relation we seek. Notice that when comparing this relation with data we assume that the Brillouin process does not contribute significantly to the peak height at $\omega = 0$; that is, the two processes must be clearly separated in frequency space. Fig. 3.8 shows that this is not strictly true for methane at $k_y = 0.12\,\text{Å}^{-1}$.

$\Delta\omega_{\mathrm{Ra}}$ is plotted for supercritical fluid methane in Fig. 3.9(a). The symbols are molecular dynamics simulation data, and the punctured line Eq. (3.71) where D_T is found from fit to data for the lowest wavevector $k_y = 0.06\,\text{Å}^{-1}$. In the low wavevector limit the classical hydrodynamic theory correctly predicts the wavevector square dependency, and $\Delta\omega_{\mathrm{Ra}}$ grows linearly with respect to the wavevector squared. It is observed that the Rayleigh peak widens for $k_y < 1\,\text{Å}^{-1}$, reaching a maximum value at approximately $1\,\text{Å}^{-1}$, and then decreases for $k_y > \text{Å}^{-1}$. The last phenomenon is referred to as de Gennes narrowing [51], and if we interpret this in terms of classical hydrodynamics must be related to a decreasing thermal diffusivity as the length scale decreases. As the length scale becomes even smaller, $\Delta\omega_{\mathrm{Ra}}$ again increases, entering a free flight limit [90].

Next, we derive the dispersion relation for the Brillouin peak. Here we will only study the peak position as a function of wavevector and derive an approximate relation. The Brillouin process, the second term in Eq. (3.67), is re-written as

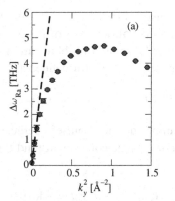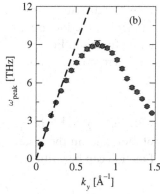

Figure 3.9 Model supercritical fluid methane at state point $(\rho, T) = (446\,\mathrm{kgm}^{-3}; 410\,\mathrm{K})$. (a) Dispersion relation for the Rayleigh half-peak width. (b) Dispersion relation for the Brillouin peak. From Ref. [91].

$$B(\mathbf{k}, \omega) = \frac{1}{2\gamma}\left[\frac{\Gamma k_y^2}{(\omega + c_s k_y)^2 + (\Gamma k_y^2)^2} + \frac{\Gamma k_y^2}{(\omega - c_s k_y)^2 + (\Gamma k_y^2)^2}\right]. \tag{3.72}$$

This is exact and can be shown by straightforward, albeit tedious, algebra. To find the maxima, we write the derivative

$$\frac{\partial B}{\partial \omega} = -\frac{\Gamma k_y^2}{\gamma}\left[\frac{c_s k_y + \omega}{((\omega + c_s k_y)^2 + (\Gamma k_y^2)^2)^2} + \frac{\omega - c_s k_y}{((\omega - c_s k_y)^2 + (\Gamma k_y^2)^2)^2}\right]$$

$$\approx -\frac{\Gamma k_y^2}{\gamma}\frac{\omega - c_s k_y}{((\omega - c_s k_y)^2 + (\Gamma k_y^2)^2)^2} \quad \text{for } \omega > 0. \tag{3.73}$$

This is zero for $\omega = c_s k_y$; hence, the peak frequency follows a linear relationship with respect to the wavevector

$$\omega_{\mathrm{peak}} = c_s k_y. \tag{3.74}$$

Figure 3.9(b) shows the dispersion relation for ω_{peak}. The prediction here (punctured line) uses the adiabatic speed of sound c_s found by Mairhofer and Sadus [152]; hence, no fitting is performed. It is seen that at low wavevector the Brillouin peak is shifted linearly with respect to wavevector, as expected. This very good agreement is not a general result. For caesium near the melting point the dispersion relation does not follow the relation using the (experimentally) measured adiabatic speed of sound [27, 90]. This has been confirmed by Bryk et al. [40], who showed that the agreement depends on the specific state-point even for supercritical fluids and liquids. The disagreement gives rise to the terms positive and negative dispersions [11], depending on whether the slope is smaller or larger than c_s. The exact length scale where classical hydrodynamic theory satisfactorily predicts the dispersion curve for the Brillouin peak is still being investigated today.

In liquid state theory the density correlations are traditionally studied through the van Hove function, Eq. (3.2), or its Fourier transform, the coherent intermediate scattering function, Eq. (3.8). We will also use the abbreviation scattering function for F; it is microscopically and for general wavevector defined as

$$F(\mathbf{k},t) = \frac{1}{N} \langle \tilde{n}(\mathbf{k},t)\tilde{n}(-\mathbf{k},0) \rangle, \tag{3.75}$$

where $\tilde{n} = \sum_i e^{-\mathbf{k}\cdot\mathbf{r}_i}$ is the number density Fourier coefficient, and N is the number of molecules in the system. For a single-component fluid the density autocorrelation function can be written as

$$C_{\rho\rho}(\mathbf{k},t) = \frac{m^2}{V} \langle \tilde{n}(\mathbf{k},t)\tilde{n}(-\mathbf{k},0) \rangle, \tag{3.76}$$

that is, we have the relation

$$F(\mathbf{k},t) = \frac{C_{\rho\rho}(\mathbf{k},t)}{m\rho_{\mathrm{av}}}. \tag{3.77}$$

The instantaneous density correlations, that is, the correlations at $t = 0$, are given by $F(\mathbf{k},0)$, or equivalently $C_{\rho\rho}(\mathbf{k},0)$. This wavevector-dependent correlation function is called the static structure factor, $S(\mathbf{k})$, and is, from Eq. (3.77), given by

$$S(\mathbf{k}) = \frac{C_{\rho\rho}(\mathbf{k},0)}{m\rho_{\mathrm{av}}}. \tag{3.78}$$

Note that if the system is isotropic, we need not be concerned with the direction of the wavevector; thus, we can simply choose $\mathbf{k} = (0,k_y,0)$ as usual. The structure factor for the model methane liquid is plotted in Fig. 3.10(a). One clearly observes that the density correlations are highly non-trivial and feature several characteristic peaks. This is simply confirming the results we saw for the van Hove function, Fig. 3.2. For small wavevectors it can be shown that the structure factor converges to a non-zero value, $\lim_{k_y\to 0} S(k_y) = \rho_{\mathrm{av}}k_B T \beta_T$ [31]; hence, the system possesses instantaneous long-ranged correlations, which is a compressibility effect.

As k_y increases, the structure factor increases and peaks at $k_y \approx 1.7$ Å^{-1}, which corresponds to approximately one molecular diameter, $2\pi/1.7$Å$^{-1} \approx 3.7$Å. In Fig. 3.10(a) it is also seen that the density correlations persist even for submolecular length scale. Obviously, these atomistic length scales are outside the realm of classical hydrodynamic theory and will not be treated in depth here; however, we do wish to address one further point.

Intuitively it is, perhaps, more informative to study the fluid structure in real space. To this end we define the radial distribution function, g, from the inverse Fourier transform of the static structure,

$$g(r) = 1 + \frac{1}{2\pi^2 r} \int_0^\infty (S(k_y) - 1) \sin(k_y r) \, dk_y, \tag{3.79}$$

where $r > 0$ and is to be understood as the radial distance to a central molecule or atom, which coordinate frame we then follow. The (average) local density surrounding this particle is related to the radial distribution function by $\rho(r) = \rho_{\mathrm{av}}g(r)$. The radial distribution function for liquid methane is shown in Fig. 3.10(b), and we clearly see a

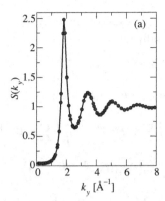
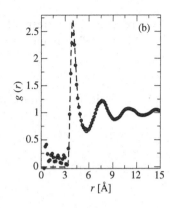

Figure 3.10 (a) Structure factor for a Lennard–Jones model of liquid methane at $(\rho, T) = (445\,\mathrm{kg m^{-3}}, 166\,\mathrm{K})$. The line connects the data and serves as a guide to the eye. (b) The corresponding radial distribution function. Filled circles result from the inverse Fourier transform definition, Eq.(3.79). The punctured line is from a direct evaluation of g.

very high density layer around one molecular diameter away from the central molecule; this is a molecular packing effect. For longer distances the radial distribution function features dampened oscillations around unity. This we interpret as the positions of the molecules further away becoming more and more uncorrelated with respect to the central one. Hence, the packing and structuring around a central molecule is less ordered. For sufficiently long distances this fluid structure is lost and g becomes unity; this long length-scale limit is the hydrodynamic regime

In our exploration of the longitudinal dynamics we have only discussed the density correlations. As has been shown, all the longitudinal hydrodynamic correlation functions depend on the same underlying physical processes, so we do not gain new insight from other longitudinal correlations. In fact, the longitudinal correlation functions are connected elegantly through the hydrodynamic relations; for example, we have that

$$\frac{\partial C_{\rho u}}{\partial t} = \frac{1}{V}\frac{\partial}{\partial t}\left\langle \widetilde{\delta\rho}(k_y,t)\widetilde{\delta u}_y(-k_y,0)\right\rangle$$
$$= \frac{1}{V}\left\langle \left(\frac{\partial}{\partial t}\widetilde{\delta\rho}(k_y,t)\right)\widetilde{\delta u}_y(-k_y,0)\right\rangle$$
$$= -ik_y\rho_{\mathrm{av}}C_{uu} \qquad (3.80)$$

from the mass balance equation. Similar relations can be found for other correlation functions (see also *Further Explorations*).

3.2 Single-Particle Dynamics

Hydrodynamics deals with collective properties; recall the microscopic definition of a hydrodynamic variable $A = A(\mathbf{r},t)$,

$$A(\mathbf{r},t) = \sum_i a_i(t)\,\delta(\mathbf{r} - \mathbf{r}_i(t)), \qquad (3.81)$$

where a_i is the corresponding microscopic quantity. Also, recall that the Fourier coefficients of A are microscopically given by

$$\widetilde{A}(\mathbf{k},t) = \sum_i a_i(t)\, e^{-i\mathbf{k}\cdot\mathbf{r}_i(t)}. \tag{3.82}$$

We can write out the autocorrelation function for A,

$$
\begin{aligned}
VC_{AA}(\mathbf{k},t) &= \left\langle \left(\sum_i a_i(t)e^{-i\mathbf{k}\cdot\mathbf{r}_i(t)}\right)\left(\sum_i a_i(0)e^{i\mathbf{k}\cdot\mathbf{r}_i(0)}\right)\right\rangle \\
&= \sum_i \left\langle a_i(t)a_i(0)e^{-i\mathbf{k}\cdot[\mathbf{r}_i(t)-\mathbf{r}_i(0)]}\right\rangle \\
&\quad + \sum_i\sum_{j\neq i}\left\langle a_i(t)a_j(0)e^{-i\mathbf{k}\cdot[\mathbf{r}_i(t)-\mathbf{r}_j(0)]}\right\rangle,
\end{aligned} \tag{3.83}
$$

using the linearity of the expected value. The first term on the right-hand side of Eq. (3.83) is a sum of single-particle correlations. The single-particle correlation is also denoted the self-part of the collective correlations. The second term in Eq. (3.83) is the cross-correlation part of the collective correlation. The cross-correlation can be very important to study; for example, it gives insight to the deviation from the Nernst–Einstein equation, which relates the self-diffusivity (single-particle process) to the charge conduction (collective property) in electrolytes [106, 107].

Based on this we define a single-particle variable as

$$A_i(\mathbf{r},t) = a_i(t)\delta(\mathbf{r}-\mathbf{r}_i), \tag{3.84}$$

with Fourier coefficients

$$\widetilde{A}_i(\mathbf{k},t) = a_i e^{-\mathbf{k}\cdot\mathbf{r}_i}. \tag{3.85}$$

From this we can evaluate the single-particle autocorrelation function using

$$\left\langle \widetilde{A}_i(\mathbf{k},t)\widetilde{A}_i(-\mathbf{k},0)\right\rangle = \frac{1}{V}\left\langle a_i(t)e^{-i\mathbf{k}\cdot\mathbf{r}_i(t)}a_i(0)e^{i\mathbf{k}\cdot\mathbf{r}_i(0)}\right\rangle. \tag{3.86}$$

As for the collective dynamics, the dynamics for the single-particle correlation are found by taking the derivative, giving

$$
\begin{aligned}
\frac{\partial\widetilde{A}_i}{\partial t} &= \left(\frac{da_i}{dt} - i\mathbf{k}\cdot\mathbf{v}_i a_i\right)e^{-i\mathbf{k}\cdot\mathbf{r}_i} \\
&\approx (1 - i\mathbf{k}\cdot\mathbf{r}_i)\frac{da_i}{dt} - i\mathbf{k}\cdot\mathbf{c}_i a_i - i\mathbf{k}\cdot\mathbf{u}(r_i,t)a_i
\end{aligned} \tag{3.87}
$$

in the low wavevector regime. We see that the dynamics of the single-particle quantities follow the same underlying dynamical structure as that of the collective dynamics.

Let us investigate the single-particle mass density ρ_i. This is defined as

$$\rho_i(\mathbf{r},t) = m_i\delta(\mathbf{r}-\mathbf{r}_i(t)), \tag{3.88}$$

or in Fourier space $\widetilde{\rho}_i(\mathbf{k},t) = m_i e^{-i\mathbf{k}\cdot\mathbf{r}_i}$. To lowest order in wavevector we have the dynamics, Eq. (3.87),

$$\frac{\partial\widetilde{\rho}_i}{\partial t} = -i\mathbf{k}\cdot m_i\mathbf{c}_i - i\mathbf{k}\cdot m_i\mathbf{u} \tag{3.89}$$

since m_i is constant. This is the low wavevector approximation of the balance equation

$$\frac{\partial \rho_i}{\partial t} = -\nabla \cdot \mathbf{J}^i - \nabla \cdot (\rho_i \mathbf{u}) , \tag{3.90}$$

where $\mathbf{J}^i$ is the single-particle flux which is due to random thermal motion of the molecules. The relevant constitutive relation here is Fick's law which, with stochastic forcing, reads

$$\mathbf{J}^i = -D_s \nabla \rho_i + \delta \mathbf{J}^i . \tag{3.91}$$

D_s is the self-diffusion coefficient and is a single-particle transport coefficient. Substituting Eq. (3.91) into Eq. (3.90), we obtain the advection-diffusion equation for ρ_i,

$$\frac{\partial \rho_i}{\partial t} = D_s \nabla^2 \rho_i - \nabla \cdot (\rho_i \mathbf{u}) - \nabla \cdot \delta \mathbf{J}^i . \tag{3.92}$$

Ignoring advection, we end up with the diffusion equation,

$$\frac{\partial \rho_i}{\partial t} = D_s \nabla^2 \rho_i - \nabla \cdot \delta \mathbf{J}^i , \tag{3.93}$$

with Fourier coefficients

$$\frac{\partial \widetilde{\rho}_i}{\partial t} = -D_s k^2 \widetilde{\rho}_i + i\mathbf{k} \cdot \widetilde{\delta \mathbf{J}}^i , \tag{3.94}$$

where $k^2 = \mathbf{k} \cdot \mathbf{k}$. Again, it is not the fluctuating quantity that we seek, but the relaxation of the correlation function. For a single-component fluid we define the single-particle correlation function

$$C_s(\mathbf{k},t) = \frac{1}{V} \langle \widetilde{\rho}_i(\mathbf{k},t) \widetilde{\rho}_i(-\mathbf{k},0) \rangle$$
$$= \frac{m^2}{V} \left\langle e^{-i\mathbf{k}\cdot\mathbf{r}_i(t)} e^{i\mathbf{k}\cdot\mathbf{r}_i(0)} \right\rangle = \frac{m^2}{V} F_s(\mathbf{k},t) , \tag{3.95}$$

where F_s is the incoherent intermediate scattering function. This is the traditional single-particle correlation function to study, and can be measured experimentally. Multiplying Eq. (3.94) by $\widetilde{\rho}_i(-\mathbf{k},0)$ and ensemble averaging over initial conditions, we arrive at the dynamical equation for F_s (or equivalently C_s):

$$\frac{\partial}{\partial t} F_s(\mathbf{k},t) = -k^2 D_s F_s(\mathbf{k},t) . \tag{3.96}$$

The general solution is, hardly surprising, an exponential function, and since

$$F_s(\mathbf{k},0) = \left\langle e^{-i\mathbf{k}\cdot\mathbf{r}(0)} e^{i\mathbf{k}\cdot\mathbf{r}(0)} \right\rangle = 1 , \tag{3.97}$$

we have the particular solution

$$F_s(\mathbf{k},t) = e^{-D_s k^2 t} . \tag{3.98}$$

This is the so-called Gaussian approximation for F_s. To understand this particular name we need to derive the corresponding real-space correlation function; this

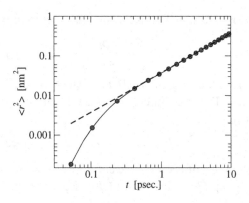

Figure 3.11   Mean square displacement (symbols connected with lines) for model liquid methane at $(\rho, T) = (446 \, \mathrm{kgm^{-3}}, 166 \, \mathrm{K})$. The punctured line represents a function proportional to time t and serves as a guide to the eye.

also helps to obtain a more physical intuitive understanding of the result. With our definition of the Fourier transform, Eq. (2.7), we get

$$G_s(\mathbf{r}, t) = \frac{1}{(2\pi)^3} \int_{-\infty}^{\infty} e^{-D_s k^2 t} \, e^{i\mathbf{k}\cdot\mathbf{r}} \mathrm{d}\mathbf{k}$$

$$= \frac{1}{8(\pi D_s t)^{3/2}} e^{-r^2/4D_s t}. \tag{3.99}$$

This is a Gaussian function and is the self-part of the van Hove correlation function G discussed in the beginning of the chapter. The term Gaussian approximation is now clear. We can interpret G_s as proportional to the conditional probability of finding the particle at position $\mathbf{r}$ at time t, if the particle was located at $\mathbf{r} = \mathbf{0}$ at $t = 0$; $\mathbf{r}$ is therefore also called the particle displacement. G_s has dimensions of inverse volume (in three dimensions) and cannot strictly be a probability distribution. Clearly, the Gaussian function is the real-space fingerprint of a diffusive process.

From the interpretation that G_s is associated with a probability distribution we can evaluate the mean, variance, and so forth of the displacement as a function of time; that is, we can evaluate the moments of G_s. The first moment is zero (i.e., the mean is zero). More interestingly, we have for the second moment, namely, the variance, $\langle r^2 \rangle$,

$$\langle r^2 \rangle = \frac{1}{8(\pi D_s t)^{3/2}} \int_{-\infty}^{\infty} r^2 e^{-r^2/4D_s t} \mathrm{d}\mathbf{r} = 6D_s t. \tag{3.100}$$

In liquid state theory $\langle r^2 \rangle$ is referred to as the mean square displacement, and we will adopt this term here. Importantly, the Gaussian approximation predicts that the mean square displacement is linear with respect to time. Figure 3.11 shows the mean square displacement of model liquid methane found from simulation. Clearly, at very short times the linear prediction fails; this is denoted the free flight or ballistic regime. For later times we see a qualitative agreement with Eq. (3.100), which we use as a definition of the Frenkel time; hence, $t > \tau_F$ for Eq. (3.100) to hold.

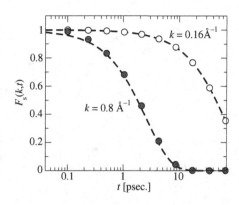

Figure 3.12 The incoherent intermediate scattering function for model methane at $(\rho, T) = (435 \text{ kgm}^{-3}, 166 \text{ K})$. Dashed lines are the predictions from Eq. (3.98) using the self-diffusion coefficient $D_s = 6.2 \times 10^{-9}$ m^{2}s^{-1}. Note, this state point is slightly different from the previous state point.

Let us return to the incoherent scattering function F_s and the theoretical predicted dynamics, Eq. (3.98). F_s can be calculated directly from the preceding definitions using equilibrium molecular dynamics simulations. The statistics can be improved considerably if we also average over all particles in the system, that is,

$$F_s(\mathbf{k},t) = \frac{1}{N}\sum_i \langle e^{-i\mathbf{k}\cdot(\mathbf{r}_i - \mathbf{r}_i(0))} \rangle. \tag{3.101}$$

As for the transverse velocity autocorrelation function, we will perform a direct comparison of the theoretical predictions to data without any fitting. To this end we need the self-diffusion coefficient D_S; this can be found from the Green–Kubo integral of the single-particle velocity autocorrelation function [156]:

$$D_s = \frac{1}{3}\int_0^\infty \langle \mathbf{c}_i(t)\cdot\mathbf{c}_i(0) \rangle \, dt. \tag{3.102}$$

We compare the data to the theory for the methane system near the triple point at two different wavevectors, Fig. 3.12. At this state point, $D_s = 6.2 \times 10^{-9}$ m^{2}s^{-1}. The predictions from the Gaussian approximation, Eq. (3.98), also are shown. Clearly, good agreement is observed. The Frenkel escape time is 2-3 psec., and the fluidic regime is reached rapidly compared to the relaxation at low wavevector. At larger wavevector the agreement is good for large times, however, the theory fails at very small time scales as expected.

This result is also observed for water at ambient conditions, see Fig. 3.13(a), where τ_F is around 7–8 psec. For the lowest wavevector the corresponding transverse velocity autocorrelation function is shown in Fig. 3.12(b). For this wavevector, $C_{uu}^\perp$ decays on a timescale much smaller than τ_F, and the relaxation does not follow simple diffusive dynamics. Thus, even if the single-particle dynamics are diffusive, the collective dynamics may not be diffusive because of cross-correlation effects.

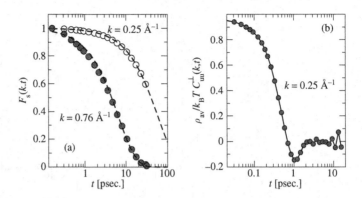

Figure 3.13 (a) The incoherent intermediate scattering function for water at ambient conditions. Dashed lines are the predictions from Eq. (3.98) using the diffusion coefficient $D_s = 2.5 \times 10^{-9}$ m^{2}s^{-1}. (b) The transverse velocity autocorrelation function.

For highly viscous systems the incoherent scattering function may feature two relaxation regimes separated by a plateau that can expand over many decays [134]. Thus, a single exponential decay predicted by Eq. (3.98) fails.

3.3 Further Explorations

1. First, verify the vector identities in Eq. (3.24). Then, use these identities and the linear constitutive relations, Eqs. (3.22), to derive the linear dynamical equations, Eqs. (3.25).
2. Derive the relation between the hydrodynamic correlation functions $C_{\rho\rho}$ and $C_{u\rho}$.
3. In Section 3.2 we treated single-particle diffusion in three dimensions. We will here study diffusion in a single dimension.

   The incoherent intermediate scattering function is still $F_s(k,t) = e^{-D_s k^2 t}$, where k is a scalar. Use the inverse Fourier transform in one dimension,

$$\mathcal{F}^{-1}[\tilde{f}(k,t)] = \frac{1}{2\pi} \int_{-\infty}^{\infty} \tilde{f}(k,t)e^{ikr}\, \mathrm{d}k,$$

   to find the self-part of the van Hove, G_s. From this show that the mean square displacement is $\langle r^2 \rangle = 2D_s t$. Use this to guess the expression for the mean square displacement in two dimensions.

   Useful integrals:

$$\int_{-\infty}^{\infty} e^{-ak^2+bk}\, \mathrm{d}k = \sqrt{\frac{\pi}{a}}\, e^{-b^2/4a} \quad (a>0, b>0)$$

$$\int_{-\infty}^{\infty} r^2 e^{-r^2/4a}\, \mathrm{d}r = 4\sqrt{\pi}a^{3/2} \quad (a>0)$$

4. From the non-advective momentum balance equation, Eq. (2.127), the shear pressure yx Fourier component can be found to be

$$\widetilde{P}_{yx}(k_y,t) = \sum_i \left(\frac{iF_{i,x}}{k_y} - m_i v_{i,x} v_{i,y} \right) e^{ik_y y_i}$$

for wavevector $\mathbf{k} = (0, k_y, 0)$. Run a molecular dynamics simulation of a Lennard–Jones liquid, see the Lennard–Jones phase diagram in Fig. 4.3, and calculate the correlation function,

$$C(k_y,t) = \frac{1}{V} \left\langle \widetilde{P}_{yx}(k_y,t) \widetilde{P}_{yx}(-k_y,0) \right\rangle .$$

Explore the behaviour of C as a function of wavevector and offer hypotheses that can explain the observations.

Computational resources available.

5. In this exploration we will see that what appears to be a simple system composed of point mass molecules can in fact feature large relaxation times. This challenges both the classic hydrodynamic description and the molecular dynamics studies of these systems due to the relatively large Deborah number.

   The system is the well-known Kob–Andersen binary Lennard–Jones mixture. The system is composed of two different particle types denoted A and B, and the particles interact in such a way that the A–B interaction is energetically favourable; this enables super-cooling while preventing phase separation. However, we shall start the exploration far away from the super-cooled regime.

   First, set the mixture ratio between A and B particles to 4:1, and let $\rho = 1.2$ and $T = 1.5$ in MD units; this corresponds to a liquid state. Estimate the pressure tensor relaxation time τ_s and viscosity η_0 from the shear pressure autocorrelation function, Eq. (3.38). From this and the Bocquet–Charlaix criterion, estimate the minimum wavelength (in MD units) where we can expect the classical picture to hold. (For reference, the standard one-component Lennard–Jones yields a wavelength of approximately 6–7σ.) You can compare your findings with the transverse velocity autocorrelation function.

   Lower the temperature in small steps of 0.1 and investigate how the minimum wavelength changes. Make sure that the system is well equilibrated and that the shear pressure autocorrelation function is fully relaxed (this will require some patience). The super-cooled regime is below $T = 1.05$; attempt to find the minimum wavelength in this region of the phase diagram.

   For further information on the Kob–Andersen fluid, see Kob and Andersen, Ref. [134], Furukawa and Tanaka, Ref. [77], and Pedersen et al., [208].

Computational resources available.

4

Extensions to Classical Hydrodynamics

In Figs. 3.4 and 3.7 we saw that the transverse velocity autocorrelation function features anti-correlations at small time and length scales. This is not predicted by the classical hydrodynamic model and thus motivates an extension to the theory. Moreover, in the classical theory the coupling between the fluid linear momentum and spin angular momentum is absent; however, in Chapter 2 we noted that in general the balance equations for these two quantities depend on the antisymmetric pressure; again, this motivates a further exploration and extension of the classical picture.

This chapter falls in two parts. In the first part we revisit the transverse velocity autocorrelation function introduced in Chapter 3. Special focus is on generalisations of Newton's law of viscosity such that short time anti-correlations and short-wavelength dynamics can be modelled in the hydrodynamic frame work.

In the second part, the hydrodynamic theory is extended to include nanoscale relaxation phenomena in molecular liquids. We then add the spin angular momentum as a new hydrodynamic variable for which the balance equation was derived in Chapter 2. Following the ideas from Chapter 3, the extended hydrodynamics is studied in detail through the relevant correlation functions in equilibrium. We also explore dielectric materials composed of permanent molecular dipoles. This is done through the polarisation density correlation function, and we will investigate the physical processes behind the relaxations of this. As a final result, we will see that the multiscale dielectric response features surprising singularities.

4.1 Viscoelastic Relaxations

The anti-correlations seen in Figs. 3.4 and 3.7 are believed to be a fingerprint of so-called shear waves and a result of an elastic component in the system's response to thermal perturbations. For water the anti-correlations are present for wavelengths on the order of nanometers and are therefore relevant for our exploration.

To highlight the phenomenon, Fig. 4.1 shows results from a DPD simulation, a simulation technique briefly introduced in Chapter 1. In what follows we write the different quantities in dimensionless DPD units, hence, we do not study any specific system and we do not give the quantities any dimensions. Note that the anti-correlation, that is, the shear waves at a given wavevector, becomes more pronounced as the temperature decreases. In the temperature range shown here, the viscosity changes by three orders

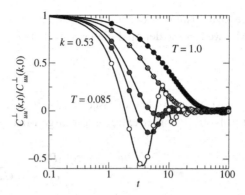

Transverse velocity autocorrelation function for a DPD system at different temperatures and fixed wavevector.

of magnitude; thus, the presence of the shear waves is correlated with high viscosity, and we refer to it as a viscous phenomenon (not saying that the increasing viscosity is the cause!).

We have seen that the classical hydrodynamic theory cannot model the super-imposed oscillations, and we will need to extend the constitutive relations that we apply to the momentum balance equation. Specifically, for the present case this means that we need a new model for the shear pressure tensor components. Recall that for the point mass atomistic systems we explore here the pressure tensor is a symmetric rank-2 tensor and we therefore only consider the trace and symmetric traceless parts. The balance equation for the fluctuations to first order is, by Eq. (3.19),

$$\rho_{\mathrm{av}}\frac{\partial \delta \mathbf{u}}{\partial t} = -\boldsymbol{\nabla} \cdot \left((p_{eq}+\Pi)\mathbf{I}+ \overset{os}{\mathbf{P}}\right),\tag{4.1}$$

and therefore the Fourier coefficients are given by

$$\rho_{\mathrm{av}}\frac{\partial \widetilde{\delta \mathbf{u}}}{\partial t} = -i\mathbf{k}\cdot\left(\widetilde{p}_{eq}+\widetilde{\Pi}\right)\mathbf{I} - i\mathbf{k}\cdot\overset{os}{\widetilde{\mathbf{P}}}.\tag{4.2}$$

As usual, we choose $\mathbf{k} = (0, k_y, 0)$ and the dynamics for the x-velocity component is

$$\rho_{\mathrm{av}}\frac{\partial \widetilde{\delta u_x}}{\partial t} = -\frac{ik_y}{2}(\widetilde{P}_{xy} + \widetilde{P}_{yx}) = -ik_y\widetilde{P}_{yx}\tag{4.3}$$

due to the symmetry of $\overset{os}{\mathbf{P}}$.

Like Newton's law of viscosity, the Maxwell constitutive model for viscoelasticity relates the gradient of the velocity field with the pressure tensor. Going back to real space, it reads in standard form, without stochastic forcing,

$$\frac{\partial u_x}{\partial y} = -\frac{1}{\eta_0}\left(1+\tau_M\frac{\partial}{\partial t}\right)P_{yx},\tag{4.4}$$

where τ_M is the Maxwell relaxation time. The Maxwell model is often formulated in terms of the stress tensor and strain rate tensor; however, we will continue to use the pressure formalism in order not to introduce unnecessary symbolism. For sufficiently small Maxwell times, denoted the viscous regime, we have that

$$\frac{\partial u_x}{\partial y} = -\frac{P_{yx}}{\eta_0}, \tag{4.5}$$

which is just Newton's law of viscosity, Eq. (3.22b). In this limit the system behaves dissipatively. In the other extreme, for large Maxwell times, that is the elastic regime, we have

$$\frac{\partial u_x}{\partial y} = -\frac{\tau_M}{\eta_0}\frac{\partial P_{yx}}{\partial t} = -\frac{1}{G_\infty}\frac{\partial P_{yx}}{\partial t}, \tag{4.6}$$

where $G_\infty = \eta_0/\tau_M$ is the modulus of rigidity. Taking this limit, we see that the shear pressure magnitude or, if you prefer, the fluid's internal stress, builds up in points, where the velocity gradient is non-zero. Equation (4.4) is a linear interpolation, or a viscoelastic model, between these two extremes.

It is worth giving a concrete example for use as a reference later. If the system undergoes a constant rate of deformation (i.e. $\partial u_x/\partial y = \dot{\gamma}_0$ everywhere and for $t \geq 0$), we have, from Eq. (4.4),

$$\frac{\partial P_{yx}}{\partial t} + \frac{1}{\tau_M}P_{yx} = -G_\infty\dot{\gamma}_0. \tag{4.7}$$

If the deformation starts from equilibrium, $P_{yx}(0) = 0$, this solves to

$$P_{yx}(t) = -\eta_0\dot{\gamma}_0\left(1 - e^{-t/\tau_M}\right). \tag{4.8}$$

In the Appendix, the method of undetermined coefficients is discussed and can be used to solve Eq. (4.7) with the given initial condition. Thus, the shear pressure relaxes at a rate determined by the Maxwell relaxation time and converges to a plateau with magnitude $\eta_0\dot{\gamma}_0$. Using the inelastic Newtonian model, the shear pressure will jump to this plateau instantaneously.

Clearly, the Maxwell model is a differential equation with respect to time for the shear pressure and is therefore an implicit expression, which cannot be applied directly to the balance Eq. (4.3). However, if the strain rate is well behaved,[1] then by the existence and uniqueness theorem [109] there exists an explicit and unique solution for P_{yx}; one such example we saw just above for constant strain rate.

With this in mind, we proceed in a more general manner and write Maxwell's model in terms of a linear differential operator, $\mathcal{A}$ [203]; that is, we can write

$$\left(1 + \tau_M\frac{\partial}{\partial t}\right)P_{yx}(t) = \mathcal{A}[P_{yx}] = -\eta_0\frac{\partial u_x}{\partial y}. \tag{4.9}$$

As just discussed, we know this equation can be solved; this we express abstractly via the operator inverse,

$$P_{yx}(t) = -\eta_0\mathcal{A}^{-1}[\partial u_x/\partial y], \tag{4.10}$$

---

[1]   This term here means Lipschitz continuous.

such that $\mathcal{A}^{-1}\mathcal{A}[P_{yx}] = \mathcal{A}\mathcal{A}^{-1}[P_{yx}] = P_{yx}$. Proving the properties of $\mathcal{A}$ and its inverse usually involves analysing the integral of the Green's function; however, we will not embark on this endeavour. Thus, by proceeding and Fourier transforming with respect to spatial coordinate, the Maxwell model reads

$$ik_y\widetilde{\delta u_x} = -\frac{1}{\eta_0}\mathcal{A}[\widetilde{P}_{yx}], \tag{4.11}$$

giving the explicit expression for the shear pressure, with stochastic forcing this time

$$\widetilde{P}_{yx} = -ik_y\eta_0\mathcal{A}^{-1}[\widetilde{\delta u_x}] - \widetilde{\delta P}_{yx}. \tag{4.12}$$

By substitution of Eq. (4.12) into Eq. (4.3), we have

$$\rho_{av}\frac{\partial\widetilde{\delta u_x}}{\partial t} = -\eta_0 k_y^2\mathcal{A}^{-1}[\widetilde{\delta u_x}] + ik_y\widetilde{\delta P}_{yx}. \tag{4.13}$$

Using the properties of the operator, this can be rearranged to give

$$\rho_{av}\left(1 + \tau_M\frac{\partial}{\partial t}\right)\frac{\partial\widetilde{\delta u_x}}{\partial t} + \eta_0 k_y^2\widetilde{\delta u_x} - ik_y\left(1 + \tau_M\frac{\partial}{\partial t}\right)\widetilde{\delta P}_{yx} = 0, \tag{4.14}$$

and collecting the terms,

$$\frac{\partial^2}{\partial t^2}\widetilde{\delta u_x} + \frac{1}{\tau_M}\frac{\partial}{\partial t}\widetilde{\delta u_x} + c_T^2 k_y^2\widetilde{\delta u_x} - ik_y\left(\tau_M\frac{\partial\widetilde{\delta P}_{yx}}{\partial t} + \widetilde{\delta P}_{yx}\right) = 0, \tag{4.15}$$

where the transverse shear wave speed,

$$c_T^2 = \eta_0/(\rho_{av}\tau_M), \tag{4.16}$$

is introduced.

Recall, we wish to derive the equation for the correlation function $C_{uu}^\perp$; thus, multiplying with $\widetilde{\delta u_x}(-k_y, 0)$ and ensemble averaging, we arrive at the differential equation

$$\frac{\partial^2}{\partial t^2}C_{uu}^\perp + \frac{1}{\tau_M}\frac{\partial}{\partial t}C_{uu}^\perp + c_T^2 k_y^2 C_{uu}^\perp = 0. \tag{4.17}$$

This is a linear second-order differential equation with constant coefficients, and the eigenvalues for this problem are found to be

$$\omega_{1,2} = -\frac{1}{2}\left(\frac{1}{\tau_M} \pm \sqrt{1/\tau_M^2 - (2c_T k_y)^2}\right). \tag{4.18}$$

In the wavevector regime $k_y > 1/(2c_T\tau_M)$, the two eigenvalues are complex (and they always come in complex conjugated pairs); and since $C_{uu}^\perp$ is a real-valued function, a property of the autocorrelation function, we have

$$C_{uu}^\perp(k_y, t) = \frac{k_B T}{\rho_{av}}e^{-t/2\tau_M}\cos(\omega_T t), \tag{4.19}$$

where the characteristic frequency is

$$\omega_T = \frac{1}{2}\sqrt{(2c_T k_y)^2 - 1/\tau_M^2} > 0. \tag{4.20}$$

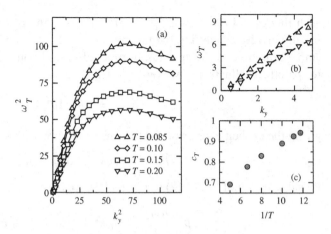

(a) and (b): Dispersion curves for ω_T for various temperatures; (b) is simply a close-up for small wave-vectors. Lines serve as a guide to the eye. It is worth noting that the Maxwell relaxation time is so large that the intersection with the y-axis is approximately zero. (c) The corresponding shear wave speeds.

This also defines the dispersion relation for the shear wave oscillations; for sufficiently large Maxwell relaxation times, the dispersion relaxation, Eq. (4.20), is to a good approximation

$$\omega_T = c_T k_y. \tag{4.21}$$

In Fig. 4.2(a) the dispersion curve for ω_T is plotted for different temperatures. We see that for low wavevectors ω_T is very low indicating a relatively large Maxwell relaxation time, thus invoking the approximation leading to Eq. (4.21). This is plotted in Fig. 4.2(b) and the slope gives the shear wave speed, Fig. 4.2(c). As can be seen, c_T increases as we approach the low-temperature viscous regime. In general, we have that the shear wave speed is smaller than the adiabatic speed of sound, $c_T < c_s$.

We have discussed the high-wavevector regime. In the low-wavevector regime, $k_y \ll 1/(2c_T\tau_M)$, the eigenvalues can be expanded to second order:

$$\omega_1 = -\eta_0 k_y^2/\rho_{\mathrm{av}} + \dots \quad \text{and} \quad \omega_2 = -1/\tau_M + \eta_0 k_y^2/\rho_{\mathrm{av}} + \dots . \tag{4.22}$$

Substituting into the general solution for Eq. (4.17), we have

$$C_{uu}^{\perp}(k_y,t) = C_1 e^{-\eta_0 k_y^2 t/\rho_{\mathrm{av}}} + C_2 e^{(\eta_0 k_y^2/\rho_{\mathrm{av}} - \tau_M^{-1})t}. \tag{4.23}$$

For zero wavevector this gives $C_{uu}^{\perp}(0,t) = C_1 + C_2 e^{-t/\tau_M}$, and therefore the integration constant C_2 must be zero since we require conservation of momentum. We then have

$$C_{uu}^{\perp}(k_y,t) = \frac{k_B T}{\rho_{\mathrm{av}}} e^{-\omega_0 t}, \tag{4.24}$$

where $\omega_0 = \eta_0 k_y^2/\rho_{\mathrm{av}}$, which is the well-known result from Chapter 3, where we ignored any elastic effects.

We can define a critical wavevector, $k_{\mathrm{crit}} = 1/(2c_T\tau_M)$, such that, if $k_y > k_{\mathrm{crit}}$, the fluid supports shear waves, but not when $k_y < k_{\mathrm{crit}}$. This is sometimes formulated as the presence of a wavevector gab, $0 \leq k_y < k_{crit}$, wherein the shear waves are not present [214]. For atomistic liquids, for example liquid argon, the critical wavelength is found to be around 0.5–1.5 nm and increases with decreasing temperature; that is, k_{crit} increases as we enter the more-viscous regime. The exact value is still debated [39].

4.1.1 Generalised Viscoelastic Modelling

A more general way to model the viscoelastic response is through a so-called transport kernel, ϕ. In this picture the momentum flux at some time t depends on the entire history of the driving force, that is, from time $t' = 0$ to $t' = t$. If $\dot{\gamma}$ is the strain rate, the shear pressure is in this formalism given by a convolution integral, excluding stochastic forcing:

$$P_{yx}(t) = -\eta_0 \int_0^t \phi(t - t') \dot{\gamma}(t') \, dt'. \tag{4.25}$$

Notice that, like the Dirac delta, the transport kernel has dimension of the inverse of the argument, here inverse of time. Treating the same case as above, we let the strain rate be zero for $t' \leq 0$ and constant $\dot{\gamma} = \dot{\gamma}_0$ for $t' > 0$; we get

$$P_{yx}(t) = -\eta_0 \dot{\gamma}_0 \int_0^t \phi(t - t') \, dt'. \tag{4.26}$$

Introducing the Maxwell kernel $\phi(t) = 1/\tau_M e^{-t/\tau_M}$, we can evaluate the convolution integral directly, giving

$$P_{yx}(t) = -\frac{\eta_0 \dot{\gamma}_0}{\tau_M} \int_0^t e^{-(t-t')/\tau_M} \, dt' = -\eta_0 \dot{\gamma}_0 (1 - e^{-t/\tau_M}), \tag{4.27}$$

which is consistent with Maxwell's constitutive model, Eq. (4.4). The Maxwell relaxation time can now be interpreted as the system characteristic memory time. In particular, we have that as $\tau_M \to 0$, the Maxwell kernel resembles a spike which we can approximate as a Dirac delta; that is, in this small time memory limit we get

$$P_{yx}(t) = -\eta_0 \dot{\gamma}_0 \int_0^t \delta(t - t') \, dt' = -\eta_0 \dot{\gamma}_0, \tag{4.28}$$

which is the Newtonian viscosity law.

Of course, we can only expect the Maxwell kernel to be applicable for certain simple fluids; a more general transport kernel can be composed of a series of Maxwell kernels with different relaxation times [174].

In the generalised formalism, we often wish to work in the frequency domain; thus, we Fourier–Laplace transform the constitutive model, using the definition from Eq. (3.41). A very important theorem that we will use a few times in what follows is the convolution theorem; this states the identity

$$\mathcal{L}\left[\int_0^\infty f(t - t') g(t') \, dt'\right] = \mathcal{L}[f(t)] \, \mathcal{L}[g(t)], \tag{4.29}$$

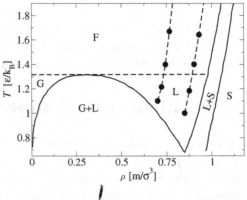

Figure 4.3 Part of the Lennard–Jones (ρ, T) phase diagram and isomorph lines (filled circles connected with punctured lines). The units are given in terms of the Lennard–Jones parameters. F, G, L, and S denote fluid, gas, liquid, and solid states, respectively.

that is, the Fourier–Laplace transform of the convolution between functions f and g equals the product of the Fourier–Laplace transforms. The convolution theorem also applies for the Fourier transform $\mathcal{F}$; we will use this below. Using the Maxwell kernel, the shear pressure in frequency domain becomes

$$\widehat{P}_{yx}(\omega) = -\frac{\eta_0 \dot{\gamma}_0}{\tau_M} \int_0^\infty e^{-(1/\tau_M + i\omega)t}\, \mathrm{d}t = -\frac{\eta_0 \dot{\gamma}_0}{1 + i\omega\tau_M} \tag{4.30}$$

at constant strain rate. It is seen that Newton's viscosity law is recaptured in the limit of zero frequency, a limit that corresponds to large times compared to the system relaxation time.

4.1.2 Hydrodynamic Invariance in Viscous Fluids

A particular type of fluid shows strong correlation between system potential energy and pressure, and is therefore named a strongly correlating system [169] or R-simple system [187]. R-simple systems feature lines in the phase diagram where dynamics and structure are invariant *if expressed in the appropriate reduced units*. These lines are called isomorphs [79]. Examples of strongly correlating systems include Lennard–Jones systems and simple united-atomic models of molecules in the dense viscous regime. Hydrogen-bonding liquids like water, on the other hand, are not strongly correlating.

Figure 4.3 shows two isomorphs superimposed on the phase diagram for the Lennard–Jones system. The isomorphs are traced using the so-called direct isomorph check iterative algorithm [79]: At some state point (ρ_1, T_1) where the system is strongly correlating the system is simulated, and the configurations $\mathbf{R}_1(t) = (\mathbf{r}_1(t), \mathbf{r}_2(t), \ldots, \mathbf{r}_N(t))$ are saved as a function of time. Each of the configurations corresponds to a potential energy $U = U(\mathbf{R}_1(t))$. $\mathbf{R}_1(t)$ is now scaled for each t, giving

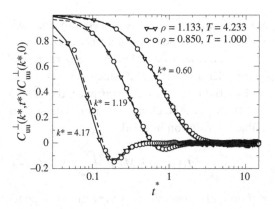

Figure 4.4 Normalised transverse velocity autocorrelation functions for two isomorphic state points. The system is a standard Lennard–Jones system. Lines serve as a guide to the eye. Data are taken from Ref. [133].

configurational vector $\mathbf{R}_2(t)$ corresponding to a new system density, ρ_2, and a potential energy, $U(\mathbf{R}_2(t))$. In order to find the new system temperature, T_2, one exploits that $U(\mathbf{R}_2)$ and $U(\mathbf{R}_1)$ are related through the correlation expression

$$U(\mathbf{R}_2) = \frac{T_2}{T_1} U(\mathbf{R}_1) + C, \tag{4.31}$$

hence, the next state point on the isomorph is (ρ_2, T_2). The algorithm is then repeated using this new state point, producing state point three on the isomorph, and so forth.

To study the invariance, the appropriate reduced units of length, time, and energy are

$$\sigma^* = n^{1/3}\sigma \quad \text{and} \quad t^* = n^{1/3}\sqrt{k_B T/mt} \quad \text{and} \quad e^* = k_B T, \tag{4.32}$$

where n is the number density and m the particle mass. In Fig. 4.4 the transverse velocity autocorrelation function is plotted for two different state points using the reduced units, Eq. (4.32). The state points are $(\rho, T) = (0.85, 1.00)$ and $(\rho, T) = (1.133, 4.233)$ and lie on the same isomorph. Note that since the Lennard–Jones fluid is only approximately invariant along an isomorph, there exist small deviations in Fig. 4.4. Importantly, the long-wavelength longitudinal dynamics are not isomorphic invariant, which is a consequence of the non-invariance of the reduced isothermal bulk modulus [133].

4.2 Non-local Viscous Response

In Sect. 4.1 the viscoelastic kernel was introduced to account for the temporal correlation effects. This idea can be extended to the spatial coordinate as well, such that, for example, the shear pressure at a given point depends on the entire strain-rate history as well as the spatial strain rate distribution in the system. Here we will revisit the transverse dynamics once more and go into detail with the kernel, focusing on the long time regime or, equivalently, small frequencies.

First, however, we will make a few general remarks, and to ease the formalism we simply consider a scalar flux J which depends on a single scalar driving force $\mathbb{X}$. For time and spatial translational invariance, the system response is independent of both the time origin and the choice of coordinate system. In this case, the flux is given by the double convolution integral,

$$J(\mathbf{r},t) = -M_0 \int_0^t \int_{-\infty}^{\infty} \phi(\mathbf{r}-\mathbf{r}',t-t')\mathbb{X}(\mathbf{r}',t')\mathrm{d}\mathbf{r}'\mathrm{d}t', \tag{4.33}$$

where M_0 is the frequency- and wavevector-independent part of the transport function. We may add a stochastic forcing term, but for these general remarks it plays no role and is omitted. If the timescale is large compared to the Maxwell relaxation time (assuming this is well defined for the system we study), we can ignore the temporal memory effects, and the system response to a driving force will only depend on the force at time t; hence, we can write the kernel as

$$\phi(\mathbf{r}-\mathbf{r}',t-t') = f(\mathbf{r}-\mathbf{r}')\delta(t-t'). \tag{4.34}$$

Note that this separation into two functions is, in general, not true. The flux then reads

$$J = -M_0 \int_0^t \delta(t-t') \int_{-\infty}^{\infty} f(\mathbf{r}-\mathbf{r}')\mathbb{X}(\mathbf{r}',t')\,\mathrm{d}\mathbf{r}'\,\mathrm{d}t'$$
$$= -M_0 \int_{-\infty}^{\infty} f(\mathbf{r}-\mathbf{r}')\mathbb{X}(\mathbf{r}',t)\,\mathrm{d}\mathbf{r}'. \tag{4.35}$$

Equivalently, if we ignore the spatial correlation effects, $f(\mathbf{r}) = \delta(\mathbf{r})$, we have

$$J = -M_0 \int_{-\infty}^{\infty} \delta(\mathbf{r}-\mathbf{r}')\mathbb{X}(\mathbf{r}',t)\,\mathrm{d}\mathbf{r}' = -M_0\mathbb{X}(\mathbf{r},t), \tag{4.36}$$

which is simply the classical local description.

From this we see that f fulfils (by design) the property

$$\int_{-\infty}^{\infty} f(\mathbf{r})\,\mathrm{d}\mathbf{r} = 1, \tag{4.37}$$

which in turn implies that

$$\widetilde{f}(\mathbf{k}) = \int_{-\infty}^{\infty} f(\mathbf{r})e^{-i\mathbf{k}\cdot\mathbf{r}}\,\mathrm{d}\mathbf{r} = 1 \text{ for } \mathbf{k} = \mathbf{0}. \tag{4.38}$$

Let us now focus on the specific case of the relaxation of the transverse autocorrelation function, $C_{uu}^{\perp}$. Again, the fundamental equation is the momentum balance equation; choosing $\mathbf{k} = (0,k_y,0)$, we have to first order in the fluctuations the streaming velocity x-component,

$$\rho_{av}\frac{\partial \widetilde{\delta u_x}}{\partial t} = -ik_y\widetilde{P}_{yx}. \tag{4.39}$$

Applying stochastic forcing to the generalised linear constitutive relation, Eq. (4.35), the shear pressure reads for sufficiently small Maxwell relaxation times

$$P_{yx}(\mathbf{r},t) = -2\eta_0 \int_{-\infty}^{\infty} f(\mathbf{r} - \mathbf{r}')\dot{\gamma}(\mathbf{r}',t)\mathrm{d}\mathbf{r}' + \delta P_{yx}, \qquad (4.40)$$

where $\dot{\gamma}$ is the yx-component of the strain-rate tensor $\overset{os}{\boldsymbol{\nabla}\mathbf{u}}$, that is,

$$\dot{\gamma} = \frac{1}{2}\left(\frac{\partial \delta u_y}{\partial x} + \frac{\partial \delta u_x}{\partial y}\right). \qquad (4.41)$$

According to the convolution theorem, we have

$$\mathcal{F}\left[\int_{-\infty}^{\infty} f(\mathbf{r} - \mathbf{r}')\dot{\gamma}(\mathbf{r}',t)\mathrm{d}\mathbf{r}'\right] = \widetilde{f}(k_y)\mathcal{F}\left[\dot{\gamma}(\mathbf{r},t)\right]$$

$$= \frac{ik_y}{2}\widetilde{f}(k_y)\widetilde{\delta u}_x(k_y,t), \qquad (4.42)$$

since we have chosen the wavevector $\mathbf{k} = (0,k_y,0)$. In Fourier space the shear pressure is then

$$\widetilde{P}_{yx} = -ik_y\eta_0\widetilde{f}(k_y)\widetilde{\delta u}_x + \widetilde{\delta P}_{yx}. \qquad (4.43)$$

Substitution of this result into Eq. (4.39) yields

$$\rho_{\mathrm{av}}\frac{\partial \widetilde{\delta u}_x}{\partial t} = -k_y^2\eta_0\widetilde{f}(k_y)\widetilde{\delta u}_x - ik_y\widetilde{\delta P}_{yx}. \qquad (4.44)$$

The dynamical equation for the transverse velocity autocorrelation function is readily obtained as

$$\rho_{\mathrm{av}}\frac{\partial}{\partial t}C_{uu}^{\perp}(k_y,t) = -k_y^2\eta_0\widetilde{f}(k_y)C_{uu}^{\perp}(k_y,t), \qquad (4.45)$$

with the solution

$$C_{uu}^{\perp}(k_y,t) = \frac{k_BT}{\rho_{\mathrm{av}}}e^{-\eta_0\widetilde{f}(k_y)k_y^2 t/\rho_{\mathrm{av}}}. \qquad (4.46)$$

This is in the same exponential form as the classical treatment, where the local constitutive relation is applied; however, the relaxation is not, in general, proportional to the wavevector squared. As for the classical treatment, Eq. (4.46) only holds for sufficiently large times, $t > \tau_M$, and will not include the shear wave phenomenon; but contrary to the classical treatment, it is valid for arbitrary wavevectors.

Equation (4.46) is not really helpful unless the function f is known; and, of course, one may propose a functional form. We here seek an expression based on data from simulations, and for this purpose we first transform Eq. (4.46) into frequency domain applying $\mathcal{L}$:

$$\widehat{C}_{uu}^{\perp}(k_y,\omega) = \frac{k_BT}{\rho_{\mathrm{av}}}\frac{1}{\frac{\eta_0\widetilde{f}(k_y)k_y^2}{\rho_{\mathrm{av}}} + i\omega}. \qquad (4.47)$$

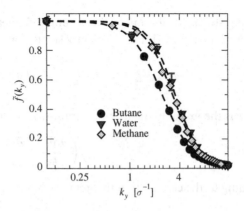

Figure 4.5 Kernel, $\widetilde{f}$, for different molecular fluids. σ is on the order of 3.1–3.9 Å. Punctured lines are best fits to Eq. (4.50). Re-plotted from Ref. [102].

Again, this only holds for sufficiently small frequencies. Especially, for $\omega = 0$ we get an expression for $\widetilde{f}$ in terms of the transverse velocity autocorrelation function when $k_y \neq 0$:

$$\eta_0 \widetilde{f}(k_y) = \frac{k_B T}{k_y^2 \widehat{C}_{uu}^{\perp}(k_y, 0)}. \tag{4.48}$$

Since we can evaluate $C_{uu}^{\perp}$ directly in simulations, this provides a very useful way to calculate $\widetilde{f}$.

Figure 4.5 plots the function $\widetilde{f}$ for liquid methane, butane, and water for different wavevector k_y. The wavevector is given in units of inverse molecular diameters, $\sigma = 3.1–3.9$ Å. It is seen that for $k_y \approx 1\sigma^{-1}$, $\widetilde{f}$ approaches the zero wavevector limit, $\widetilde{f} \to 1$. This indicates that the local description (i.e., Newton's law of viscosity) holds down to a few nanometres for these simple systems.

There are no satisfactory theories that can predict the kernels from first principles; for example, standard mode coupling theory fails [77]. Different empirical functional forms have been suggested, for example,

$$\widetilde{f}(k_y) = \frac{1}{1 + (\gamma k_y)^2 + (\xi k_y)^4} \tag{4.49}$$

by Furukawa and Tanaka [77], and

$$\widetilde{f}(k_y) = \frac{1}{1 + \alpha k_y^{\beta}} \tag{4.50}$$

by Hansen et al. [94]. The punctured lines in Fig. 4.5 represent best fit of Eq. (4.50) to data.

From these fits one can define a transverse dynamical length scale,

$$L^{\perp} = \alpha^{1/\beta}. \tag{4.51}$$

This quantifies a length where the collective spatial correlations in the liquid affects the viscosity. For non-viscous fluids $L^{\perp}$ is usually on the order of around one molecular

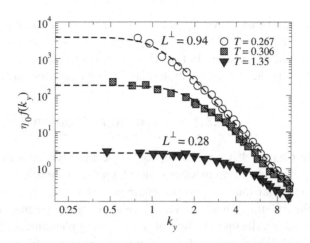

Figure 4.6 Viscosity kernels for a binary Lennard–Jones system at different temperatures. At temperature $T = 1.35$, the system is in the non-viscous liquid state, and at $T = 0.267$, the system is in the super-cooled viscous state. The lines represent best fit to Eq. (4.50) from which $L^\perp$ is calculated. The units are reduced molecular dynamics units. Data are taken from Furukawa and Tanaka [77].

diameter or less, but it increases as the system approaches the viscous regime [77, 104]. Figure 4.6 shows the wavevector-dependent zero frequency viscosity $\eta(k_y) = \eta_0 f(k_y)$ for a two-component Lennard–Jones system at three different temperatures. The quantities are given in MD units and not written explicitly. The binary mixture allows the system to be super-cooled and feature very slow dynamics without crystallising. The result shows that, on approaching the viscous regime, the viscosity η_0 increases dramatically; but the length scale $L^\perp$ does not increase with the same rate, meaning that the spatial correlations are not affected in the same dramatic manner. This indicates that it is not an increase in correlation length that is the physics behind the increasing viscous behaviour in the super-cooled regime of simple liquids. For polymer melts approaching the glassy state, on the other hand, molecular dynamics simulations have indicated that $L^\perp$ diverges [180]; hence, the correlation length is still not fully understood.

The generalisation presented here for the transverse dynamics can also be applied to the longitudinal dynamics, leading to kernels for, say, the bulk viscosity and heat conductivity. A more heuristic approach is to simply fit the classical hydrodynamic correlation functions to data allowing thermal diffusivity, attenuation coefficient, and adiabatic speed of sound to be wavevector dependent. We will not pursue this further here.

4.3 Relaxation Phenomena in Molecular Fluids

We have mainly studied nanoscale fluid systems, where the constituent molecules are point mass particles; the workhorse methane is an example of such a point mass system. The hydrodynamic correlation functions are also applied to study the hydrodynamics of more complex fluids like water; see Bertolini and Tani, Ref. [19], for a very comprehensive investigation. However, the theory developed in Chapter 2 allows us to extend the exploration by including the effects of molecular rotation, that is, molecular degrees of freedom. In fact, this extension will later be critical in order to understand certain phenomena for molecular flows on the nanoscale.

4.3.1 Spin Angular Momentum Coupling

In Chapter 1 it was shown from molecular dynamics simulations that a flow can be achieved by rotating water molecules using an electric field. This coupling between the molecular rotation and the streaming velocity cannot be explained from the classical theory, and we must therefore extend the hydrodynamic model. Inspired by Chapter 3, we begin the investigation of this coupling phenomenon in equilibrium, and we shall derive the relevant hydrodynamic correlation functions.

First, recall that in our molecular formalism the pressure is not in general a symmetric tensor, and that the antisymmetric part appears in both the equation for the streaming velocity and the spin angular momentum; see Eqs. (2.62) and (2.113). It is this fact that leads to the coupling between the two hydrodynamic variables.

In equilibrium the average angular velocity is zero, $\mathbf{\Omega}_{av} = 0$; hence, $\mathbf{\Omega} = \delta\mathbf{\Omega}$ and also we have $\mathbf{u} = \delta\mathbf{u}$. Then the balance equations for the fluctuating parts $\delta\mathbf{u}$ and $\delta\mathbf{\Omega}$ take the following general forms:

$$\rho_{av}\frac{\partial\delta\mathbf{u}}{\partial t} = -\mathbf{\nabla}\cdot((p_{eq}+\Pi)\mathbf{I}+\overset{os}{\mathbf{P}}) - \mathbf{\nabla}\times\overset{ad}{\mathbf{P}} \tag{4.52a}$$

$$\rho_{av}\Theta\frac{\partial\delta\mathbf{\Omega}}{\partial t} = -2\overset{ad}{\mathbf{P}} -\mathbf{\nabla}\cdot(Q+\overset{os}{\mathbf{Q}}) - \mathbf{\nabla}\times\overset{ad}{\mathbf{Q}}. \tag{4.52b}$$

We have already introduced the constitutive relations for the viscous normal pressure Π and the traceless symmetric part of the pressure tensor $\overset{os}{\mathbf{P}}$ in Eqs. (3.22). To proceed, we need four additional constitutive relations to account for the fluxes $\overset{ad}{\mathbf{P}}, Q, \overset{os}{\mathbf{Q}}$, and $\overset{ad}{\mathbf{Q}}$.

We begin with the pseudo-vector $\overset{ad}{\mathbf{P}}$; to this end, we need to discuss the motion of a rigid body. Microscopically, for a rigid-body fluid element the molecules, on average, maintain their relative positions; this is denoted local molecular rigidity [67]. Now, from Eq. (2.59) we saw that there exists a torque due to the non-symmetric molecular interactions, and this torque will lead to a molecular rotation and, hence, a deviation from local rigidity. To propose a constitutive relation for this resulting antisymmetric pressure, we first need a result for the local rotation of a rigid body. For simplicity, consider a fluid that flows in a circular path around a reference point $\mathbf{0}$, see Fig. 4.7, and note that both the orbital angular velocity and the spin angular velocity are given by the spatially constant angular velocity $\mathbf{\Omega}$. Moreover, at any point $\mathbf{r}$ the local streaming velocity is $\mathbf{u} = \mathbf{\Omega}\times\mathbf{r}$, and since the local rotation is given by the vorticity we get

$$\mathbf{\nabla}\times\mathbf{u} = \mathbf{\nabla}\times(\mathbf{\Omega}\times\mathbf{r})$$
$$= \mathbf{\Omega}(\mathbf{\nabla}\cdot\mathbf{r}) - (\mathbf{\Omega}\cdot\mathbf{\nabla})\mathbf{r} = 2\mathbf{\Omega}. \tag{4.53}$$

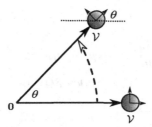

Figure 4.7 Illustration of a rigid body flow. The circles represent the fluid element $\mathcal{V}$ as it undergoes a rigid-body rotation. Superimposed arrows indicate the local rotation of $\mathcal{V}$ as it flows, indicated by the large arrow.

Thus, for a rigid body rotation, the local spin angular velocity is the same at every point and given by the velocity field alone.

The deviation from local molecular rigidity is given by the difference between the vorticity and twice the angular velocity, that is,

$$\overset{ad}{\mathbf{P}} = -\eta_r(\boldsymbol{\nabla} \times \mathbf{u} - 2\boldsymbol{\Omega}) + \delta \overset{ad}{\mathbf{P}}, \tag{4.54}$$

where η_r is the rotational viscosity. Notice that Curie's principle is fulfilled.

The spin angular momentum flux tensors $Q, \overset{os}{\mathbf{Q}}$, and $\overset{ad}{\mathbf{Q}}$ we model in the standard manner through gradients in the spin angular velocity [52, 67]:

$$Q = -\zeta_v(\boldsymbol{\nabla} \cdot \boldsymbol{\Omega}) + \delta Q \tag{4.55a}$$

$$\overset{os}{\mathbf{Q}} = -\zeta_0 \left[(\boldsymbol{\nabla}\boldsymbol{\Omega} + \boldsymbol{\Omega}\boldsymbol{\nabla}) - \frac{2}{3}\mathrm{Tr}(\boldsymbol{\nabla}\cdot\boldsymbol{\Omega})\mathbf{I} \right] + \delta\overset{os}{\mathbf{Q}} \tag{4.55b}$$

$$\overset{ad}{\mathbf{Q}} = -\zeta_r(\boldsymbol{\nabla} \times \boldsymbol{\Omega}) + \delta\overset{ad}{\mathbf{Q}}. \tag{4.55c}$$

Here ζ_0, ζ_r, and ζ_v are called the shear, rotational, and bulk spin viscosities.

Substitution of the constitutive relations into Eqs. (4.52) gives

$$\rho_{\mathrm{av}}\frac{\partial \delta\mathbf{u}}{\partial t} = -\boldsymbol{\nabla}p_{eq} + (\eta_v + \eta_0/3 - \eta_r)\boldsymbol{\nabla}(\boldsymbol{\nabla}\cdot\delta\mathbf{u}) + (\eta_0 + \eta_r)\nabla^2\delta\mathbf{u}$$
$$+ 2\eta_r\boldsymbol{\nabla} \times \delta\boldsymbol{\Omega} + \boldsymbol{\nabla}\cdot\delta\mathbf{P} \tag{4.56a}$$

$$\rho_{\mathrm{av}}\Theta\frac{\partial \delta\boldsymbol{\Omega}}{\partial t} = 2\eta_r(\boldsymbol{\nabla}\times\delta\mathbf{u} - 2\delta\boldsymbol{\Omega}) - (\zeta_v + \zeta_0/3 - \zeta_r)\boldsymbol{\nabla}(\boldsymbol{\nabla}\cdot\delta\boldsymbol{\Omega})$$
$$- (\zeta_0 + \zeta_r)\nabla^2\delta\boldsymbol{\Omega} + \boldsymbol{\nabla}\cdot\delta\mathbf{Q} + 2\delta\overset{ad}{\mathbf{P}}, \tag{4.56b}$$

where $\delta\mathbf{P} = \delta\Pi\mathbf{I} + \delta\overset{os}{\mathbf{P}} + \delta\overset{a}{\mathbf{P}}$ and $\delta\mathbf{Q} = \delta Q\mathbf{I} + \delta\overset{os}{\mathbf{Q}} + \delta\overset{a}{\mathbf{Q}}$. For ease of reading, the transverse viscosity, η_t, and the transverse spin viscosity, ζ_t, are introduced in the following:

$$\eta_t = \eta_0 + \eta_r, \text{ and } \zeta_t = \zeta_0 + \zeta_r. \tag{4.57}$$

In terms of the Fourier coefficients, we get the dynamics

$$\rho_{av}\frac{\partial \widetilde{\delta \mathbf{u}}}{\partial t} = -i\mathbf{k}\widetilde{p}_{eq} - (\eta_v + \eta_0/3 - \eta_r)\mathbf{k}(\mathbf{k}\cdot\widetilde{\delta \mathbf{u}}) - \eta_t k^2 \widetilde{\delta \mathbf{u}}$$

$$+ 2i\eta_r \mathbf{k}\times\widetilde{\delta \mathbf{\Omega}} + i\mathbf{k}\cdot\widetilde{\delta \mathbf{P}} \tag{4.58a}$$

$$\rho_{av}\Theta\frac{\partial \widetilde{\delta \mathbf{\Omega}}}{\partial t} = 2\eta_r(i\mathbf{k}\times\widetilde{\delta \mathbf{u}} - 2\widetilde{\delta \mathbf{\Omega}}) - (\zeta_v + \zeta_0/3 - \zeta_r)\mathbf{k}(\mathbf{k}\cdot\widetilde{\delta \mathbf{\Omega}}) - \zeta_t k^2 \widetilde{\delta \mathbf{\Omega}}$$

$$+ i\mathbf{k}\cdot\widetilde{\delta \mathbf{Q}} + 2\widetilde{\delta\,\overset{ad}{\mathbf{P}}}. \tag{4.58b}$$

We have the wavevector $\mathbf{k} = (0, k_y, 0)$, and we first focus on the longitudinal spin angular dynamics, as this does not couple to other hydrodynamic variables. From Eq. (4.58b) we see that

$$\rho_{av}\Theta\frac{\partial \widetilde{\delta\Omega_y}}{\partial t} = -(4\eta_r + \zeta_l k_y^2)\widetilde{\delta\Omega_y} + ik_y\widetilde{\delta Q_{yy}} + 2\widetilde{\delta\,\overset{ad}{P}_y}, \tag{4.59}$$

where $\zeta_l = \zeta_v + 4\zeta_0/3$ is longitudinal spin viscosity. We can now construct the longitudinal spin angular velocity autocorrelation function as

$$C_{\Omega\Omega}^{\parallel}(k_y, t) = \frac{1}{V}\langle \widetilde{\delta\Omega_y}(k_y, t)\widetilde{\delta\Omega_y}(-k_y, 0)\rangle, \tag{4.60}$$

and the dynamical equation for $C_{\Omega\Omega}^{\parallel}$ is given by multiplication of $\widetilde{\delta\Omega_y}(-k_y, 0)$ and ensemble averaging Eq. (4.59). One obtains

$$\rho_{av}\Theta\frac{\partial C_{\Omega\Omega}^{\parallel}}{\partial t} = -(4\eta_r + \zeta_l k_y^2)C_{\Omega\Omega}^{\parallel}. \tag{4.61}$$

By applying the equipartition theorem, it can be shown that the initial condition is [102]

$$C_{\Omega\Omega}^{\parallel}(k_y, 0) = \frac{9k_B T}{4\rho_{av}\Theta}, \tag{4.62}$$

and therefore the particular solution to Eq. (4.61) reads

$$C_{\Omega\Omega}^{\parallel}(k_y, t) = \frac{9k_B T}{4\rho_{av}\Theta}e^{-\omega t}. \tag{4.63}$$

The eigenvalue is given by

$$\omega = \frac{4\eta_r + \zeta_l k_y^2}{\rho_{av}\Theta}, \tag{4.64}$$

which also defines the relevant dispersion relation.

To compare the predictions from this extended hydrodynamic model with simulation data without fitting, we need to calculate the transport coefficients η_r and ζ_l. Evans and Hanley [64] argued that, unlike the shear viscosity, the rotational viscosity η_r does not have a standard Green–Kubo integral, as it describes a wavevector-independent process. Evans and Hanley also showed that it can be evaluated from the generalised Langevin equation; however, due to statistical noise in the data, the method can be significantly improved by introducing an empirical relation between the antisymmetric stress autocorrelation function, $\overset{a}{C}$, and rotational viscosity

$$\overset{a}{C}(s) = \frac{\eta_r s}{4\eta_r/\rho_{av}\Theta + s + s^2\tau},$$ (4.65)

where τ is a relaxation time, s is the Laplace coordinate, and

$$\overset{a}{C}(s) = \frac{V}{3k_BT}\int_0^\infty \left\langle \overset{ad}{\mathbf{P}}(t) \cdot \overset{ad}{\mathbf{P}}(0) \right\rangle e^{-st}\, dt.$$ (4.66)

The pseudo-vector $\overset{ad}{\mathbf{P}}$ is defined microscopically in Eq. (2.59), the correlation function $\overset{a}{C}(s)$ can be calculated from Eq. (4.66), and η_r is found from fitting to Eq. (4.65).

We also need the spin viscosity, ζ_l. From simulations it has been shown that the spin viscosities ζ_l and ζ_t are equal within statistical uncertainty, at least in the limit of small wavevector and small moment of inertia [101, 98], that is, $\zeta_l \approx \zeta_t$. We will therefore from here on use the same symbol for both the spin viscosities, namely ζ. The spin viscosity does have an ordinary Green–Kubo integral [67]:

$$\zeta = \frac{V}{2k_BT}\int_0^\infty \left\langle \overset{os}{Q}_{xy}(t)\, \overset{os}{Q}_{xy}(0) + \overset{a}{Q}_{xy}(t)\, \overset{a}{Q}_{xy}(0) \right\rangle dt.$$ (4.67)

The computed values for η_r and ζ for chlorine, butane, and water are listed in Table 4.1.

The correlation function $C_{\Omega\Omega}^{\parallel}$ is computed from the microscopic definition of the spin angular velocity; to first order in the fluctuations we have

$$\widetilde{\delta\Omega} \approx \frac{1}{\rho_{av}\Theta}\sum_i \mathbf{S}_i e^{-\mathbf{k}\cdot\mathbf{r}_i(t)},$$ (4.68)

where $\mathbf{S}_i$ is the angular momentum of molecule i, Eq. (2.101). Figure 4.8(a) plots the spectrum, specifically, the imaginary part of the correlation function $C_{\Omega\Omega}^{\parallel}$ for a generic diatomic molecular fluid. For zero wavevector the dispersion relation, Eq. (4.64), reveals a characteristic frequency, ω_c:

$$\omega_c = \frac{4\eta_r}{\rho_{av}\Theta}.$$ (4.69)

This frequency is indicated by the vertical punctured line in Fig. 4.8(a) using the values for η_r found from Eq. (4.65). Note, for zero wavevector the relaxation of the spin angular momentum is governed by the coupling to the linear momentum alone.

For non-zero wave-vectors the peak frequency increases with increasing wavevector. The underlying process for this behaviour is diffusion of spin angular momentum, as predicted by the theory. In Fig. 4.8(b) the corresponding dispersion plot is shown. The dashed line plots the predictions from the theory and is in good agreement with the simulation data. How well the theory performs for other molecular liquids is still not known in detail. In the last column of Table 4.1 the characteristic frequency is given based on the values for the viscosities. Clearly, this relaxation is an extremely fast process for small-weight molecular fluids.

Table 4.1 Rotational and spin viscosities for different liquids. The last column lists the characteristic angular spin relaxation time. From Refs. [160, 92, 102]. ^{a} state point: 1,605 kgm^{-3}, 194.0 K, ^{b} state point: 582.3 kgm^{-3}, 288.0 K, ^{c} state point: 996.3 kgm^{-3}, 298.7 K

Molecule	η_0 [mPa·s]	η_r [mPa·s]	ζ [kg ms^{-1}]	ω_c [THz]
Chlorine^a	0.74	0.47	8.3×10^{-24}	166
Butane^b	0.18	0.013	4.0×10^{-24}	7.4
Water^c	0.75	0.17	2.1×10^{-21}	812

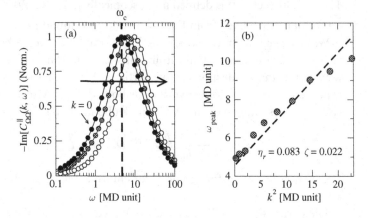

Figure 4.8 (a) The spectrum for $C_{\Omega\Omega}^{\parallel}$ for three different wavevectors. Lines serve as a guide to the eye, and the arrow indicates increasing wavevector. (b) The corresponding dispersion relation. The dashed line is the theoretical prediction; values for η_r and ζ are calculated via Eqs. (4.65) and (4.67). The system is a generic diatomic molecule at a supercritical fluidic state; data are re-plotted from Ref. [102].

We now return to the transverse dynamics. From Eqs. (4.58a) and (4.58b) it can be shown that

$$\rho_{av}\frac{\partial \widetilde{\delta u_x}}{\partial t} = -\eta_t k_y^2 \widetilde{\delta u_x} + 2i\eta_r k_y \widetilde{\delta\Omega_z} - ik_y \widetilde{\delta P_{yx}} \tag{4.70a}$$

$$\rho_{av}\Theta\frac{\partial \widetilde{\delta\Omega_z}}{\partial t} = -2i\eta_r k_y \widetilde{\delta u_x} - (4\eta_r + \zeta k_y^2)\widetilde{\delta\Omega_z} - ik_y \widetilde{\delta Q_{yz}} - 2\widetilde{\delta P_z^{ad}}. \tag{4.70b}$$

Thus, the dynamics of the velocity x-component and the dynamics of the angular velocity z-component are coupled. Following the procedure in Chapter 3, we can define four correlation functions by multiplying Eqs. (4.70) by $\widetilde{\delta u_x}(-k_y, 0)$ or $\widetilde{\delta\Omega_z}(-k_y, 0)$ and taking the average over an ensemble of independent initial conditions. Similar to Eq. (3.53), we obtain the matrix system of correlation functions

$$\frac{\partial}{\partial t}\begin{bmatrix} C_{uu}^{\perp} & C_{u\Omega}^{\perp} \\ C_{\Omega u}^{\perp} & C_{\Omega\Omega}^{\perp} \end{bmatrix} = \begin{bmatrix} -\dfrac{\eta_t k_y^2}{\rho_{av}} & \dfrac{i2\eta_r k_y}{\rho_{av}} \\ -\dfrac{i2\eta_r k_y}{\rho_{av}\Theta} & -\dfrac{4\eta_r + \zeta k_y^2}{\rho_{av}\Theta} \end{bmatrix} \cdot \begin{bmatrix} C_{uu}^{\perp} & C_{u\Omega}^{\perp} \\ C_{\Omega u}^{\perp} & C_{\Omega\Omega}^{\perp} \end{bmatrix}. \tag{4.71}$$

From this we see that the dynamics can be written as two sets of co-dependent correlation functions,

$$\{C_{uu}^{\perp}, C_{\Omega u}^{\perp}\} \text{ and } \{C_{\Omega\Omega}^{\perp}, C_{u\Omega}^{\perp}\}. \tag{4.72}$$

Here we will simply study the first set. The eigenvalues found from the 2-by-2 hydrodynamic matrix are

$$\omega_{1,2} = \frac{4\eta_r + (\zeta + \eta_t\Theta)k_y^2 \pm \sqrt{d}}{2\rho_{\text{av}}\Theta}, \tag{4.73}$$

where the discriminant is

$$d = 16\eta_r^2 + (8\eta_r\zeta - 16\eta_r^2\Theta + 8\eta_r\eta_t\Theta)k_y^2 + (\zeta^2 - 2\eta_t\zeta\Theta + \eta_t^2\Theta^2)k_y^4. \tag{4.74}$$

Using the initial conditions

$$C_{uu}^{\perp}(k_y, 0) = k_B T/\rho_{\text{av}} \text{ and } C_{\Omega u}^{\perp}(k_y, 0) = 0, \tag{4.75}$$

we obtain the particular solution [102]

$$C_{uu}^{\perp}(k_y, t) = \frac{k_B T}{2\rho_{\text{av}}\sqrt{d}} \left((\sqrt{d} - A)e^{-\omega_1 t} + (\sqrt{d} + A)e^{-\omega_2 t} \right) \tag{4.76a}$$

$$C_{\Omega u}^{\perp}(k_y, t) = i\frac{2k_B T \eta_r k_y}{\rho_{\text{av}}\sqrt{d}}(e^{-\omega_1 t} - e^{-\omega_2 t}), \tag{4.76b}$$

with $A = 4\eta_r + (\zeta - \eta_t\Theta)k_y^2$.

Before we compare this result with simulation data, it is informative to study the small wavevector limit. Expansion of the eigenvalues yields to second order in wavevector

$$\omega_1 = \frac{1}{\rho_{\text{av}}\Theta} \left(4\eta_r + (\zeta + \eta_r\Theta)k_y^2 \right) + \dots \tag{4.77a}$$

$$\omega_2 = \frac{\eta_0 k_y^2}{\rho_{\text{av}}} + \dots, \tag{4.77b}$$

and for d we have

$$\sqrt{d} = 4\eta_r + (\zeta + 2\eta_r\Theta - \eta_t\Theta)k_y^2 + \dots. \tag{4.78}$$

From the prefactor in Eq. (4.76b) we then have that

$$C_{\Omega u}^{\perp}(k_y, t) \to 0 \text{ as } k_y \to 0, \tag{4.79}$$

that is, the coupling between the spin angular momentum and the streaming velocity vanishes in the small wavevector limit, that is, in the classical hydrodynamic regime.

In this limit we also have that the prefactor for the first term in the transverse velocity autocorrelation function, Eq. (4.76a), is

$$\sqrt{d} - A \approx 2\eta_r\Theta k_y^2. \tag{4.80}$$

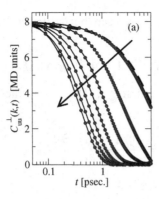
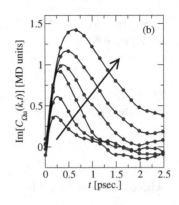

Figure 4.9   Model molecular chlorine in the supercritical fluidic state $(\rho, T) = (795.5 \text{ kgm}^{-3}, 713.0 \text{ K})$. (a) Transverse velocity autocorrelation function, $C_{uu}^{\perp}$. (b) The cross-correlation function $C_{\Omega u}^{\perp}$. In (a) the punctured line represents the classical hydrodynamic prediction using η_0=0.077 mPa ·s. Arrows indicate increasing wavevector in the interval $0.1 \text{ Å}^{-1} \leq k_y \leq 0.7 \text{Å}^{-1}$. Data are re-plotted from Ref. [101].

Since Θ is typically in the order of 10^{-20} m^2 for small molecules like chlorine and butane, and η_r is of the same order as η_0, the first term in Eq. (4.76a) can safely be ignored and we recapture the classical result from Chapter 3,

$$C_{uu}^{\perp}(k_y, t) = \frac{k_B T}{\rho_{av}} e^{-\omega_0 t} \quad (\text{small } k_y \text{ and small } \Theta), \qquad (4.81)$$

where $\omega_0 = \eta_0 k^2 / \rho_{av}$. The fact that the coupling vanishes in the classical hydrodynamic regime and we recapture the classical relaxation result for the transverse velocity auto-correlation function in the low wavevector limit implies that this coupling phenomena is a true nanoscale phenomenon.

Figure 4.9(a) shows molecular dynamics data for the transverse velocity autocorre-lation function $C_{uu}^{\perp}$ for molecular chlorine fluid. Note that the classical hydrodynamic prediction is recaptured at low wavevectors expected from the preceding discussion. In Fig. 4.9(b) $C_{\Omega u}^{\perp}$ is shown and the prediction from the theory, Eq. (4.79), is confirmed, as the coupling between the spin angular velocity and the streaming velocity vanishes in the small wavevector limit, whereas it increases for increasing wavevector.

We finish this section with an important observation. The spin angular velocity does not couple to the density and thermal kinetic energy, as pointed out by Evans and Streett [67]. The hydrodynamic exploration of the Rayleigh and Brillouin processes discussed in Chapter 3 need not to be extended for molecular systems.

4.4 Polarisation Relaxation

Recall from Chapter 2 that the polarisation **P** can microscopically be defined as [90]:

$$\mathbf{P}(\mathbf{r}, t) = \rho(\mathbf{r}, t)\mathbf{p}(\mathbf{r}, t) = \sum_i \boldsymbol{\mu}_i(t)\delta(\mathbf{r} - \mathbf{r}_i), \qquad (4.82)$$

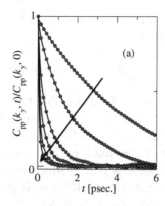
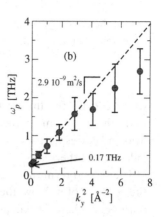

Figure 4.10 Molecular dynamics simulation results for water at ambient conditions. (a) The dipole moment autocorrelation function for different wavevectors. Arrow indicates increasing wavevector. (b) Corresponding dispersion plot for the peak frequency. Dashed line represents best fit of $\omega_p = 1/\tau_D + \nu k_y^2$, to data (the values indicate the fitted values). From [100].

where **p** is the dipole moment per unit mass, and $\boldsymbol{\mu}$ the microscopic molecular dipole. It was noted earlier that this definition of the polarisation is, in fact, an approximation and, as promised in Chapter 2, we will now return to this point.

Let $\mathbf{p} = \mathbf{p}_{av} + \delta\mathbf{p}$, where in equilibrium $\mathbf{p}_{av} = \mathbf{0}$ and we have to first order in the fluctuations

$$\delta\mathbf{p}(\mathbf{r},t) = \frac{1}{\rho_{av}}\sum_i \boldsymbol{\mu}_i(t)\delta(\mathbf{r}-\mathbf{r}_i). \tag{4.83}$$

The Fourier coefficients follow directly:

$$\widetilde{\delta\mathbf{p}}(\mathbf{k},t) = \frac{1}{\rho_{av}}\sum_i \boldsymbol{\mu}_i(t)e^{-\mathbf{k}\cdot\mathbf{r}_i}. \tag{4.84}$$

For $\mathbf{k} = (0,k_y,0)$ we define the dipole moment autocorrelation function, C_{pp}, as

$$C_{pp}(k_y,t) = \frac{1}{V}\left\langle \delta p_y(k_y,t)\delta p_y(-k_y,0)\right\rangle, \tag{4.85}$$

where δp_y is the dipole moment y-component. This correlation function is the fundamental function when studying the polarisation relaxation dynamics.

Figure 4.10(a) shows simulation results for the dipole moment autocorrelation function C_{pp} for water at ambient conditions, and Fig. 4.10(b) shows the corresponding dispersion relation for the peak frequency. From the dispersion relation one observes

1. a relaxation process at zero wavevector, and
2. a relaxation process depending on the wavevector squared, that is, a diffusive process.

To understand this dispersion relation, we first recall the balance equation for the polarisation, Eq. (2.126). In equilibrium and to first order in the fluctuations we can ignore the cross coupling with the angular velocity, the advective term, and, of course, the external electric field, that is,

$$\rho_{av}\frac{\partial \delta \mathbf{p}}{\partial t} = \frac{1}{\tau_D}\rho_{av}\delta \mathbf{p} - \mathbf{\nabla} \cdot \mathbf{R} - \mathbf{\nabla} \cdot \delta \mathbf{R}, \tag{4.86}$$

where $\delta \mathbf{R}$ are the fluctuating part of the dipole flux tensor. The Debye relaxation term, the first term on the right-hand side, is kept even in the absence of an electric field and corresponds to a thermal limit where the molecular dipoles are considered as independent thermally relaxing entities; this is referred to as the Debye limit.

The second term represents a diffusive process which we expect to be dependent on the gradient of $\mathbf{p}$. Now, the dipole flux tensor $\mathbf{R}$ is not, in general, symmetric, but for our purpose here we will assume that the antisymmetric part can be ignored, and the tensor is decomposed into

$$\mathbf{R} = R\mathbf{I} + \overset{os}{\mathbf{R}}. \tag{4.87}$$

R is one third of the trace, and $\overset{os}{\mathbf{R}}$ the traceless symmetric part of $\mathbf{R}$, respectively. Following the idea from Chapter 3 these fluxes are modelled from the local linear constitutive relations [99]:

$$R = -\chi_v(\mathbf{\nabla} \cdot \mathbf{p}) + \delta R \tag{4.88a}$$

$$\overset{os}{\mathbf{R}} = -2\chi_0 \overset{os}{(\mathbf{\nabla}\mathbf{p})} + \delta\overset{os}{\mathbf{R}}. \tag{4.88b}$$

Here χ_v and χ_0 are the associated linear transport coefficients. Substituting and Fourier transforming, we end up with the dynamical equation for the Fourier coefficients,

$$\rho_{av}\frac{\partial \widetilde{\delta \mathbf{p}}}{\partial t} = -\frac{\rho_{av}}{\tau_D}\widetilde{\delta \mathbf{p}} - \chi_l\mathbf{k}(\mathbf{k} \cdot \widetilde{\delta \mathbf{p}}) - \chi_0 k^2 \widetilde{\delta \mathbf{p}} - i\mathbf{k} \cdot \widetilde{\delta \mathbf{R}}, \tag{4.89}$$

where $\widetilde{\delta \mathbf{R}}$ represents the total stochastic forcing and $\chi_l = \chi_v + 4\chi_0/3$. From this we can form the dynamical equation for the dipole moment autocorrelation functions, Eq. (4.85). Since $\mathbf{k} = (0, k_y, 0)$ and we obtain

$$\frac{\partial \widetilde{\delta p_y}}{\partial t} = -\left(\frac{1}{\tau_D} + (v_0 + v_l)k_y^2\right)\widetilde{\delta p_y} - \frac{ik_y}{\rho_{av}}\widetilde{\delta R_{yy}}, \tag{4.90}$$

using the definitions for the kinematic transport coefficients $v_0 = \chi_0/\rho_{av}$ and $v_l = \chi_l/\rho_{av}$. We multiply Eq. (4.90) with $\delta p_y(-k_y, 0)$, ensemble averaging, and solve the differential equation, giving the usual exponential relaxation,

$$C_{pp}(k_y, t) = C_{pp}(k_y, 0)e^{-\omega_p t}, \tag{4.91}$$

with

$$\omega_p = \frac{1}{\tau_D} + (v_t + v_l)k_y^2. \tag{4.92}$$

This is exactly the form of the dispersion relation we expected from the simulation data. If we compare with Fig. 4.10, we obtain a Debye relaxation time of approximately

6 psec. and a kinematic transport coefficient $v = v_t + v_l = 2.9 \times 10^{-9}$ m^{2}s^{-1} for the flexible simple point charge water model (SPC/Fw).

For confined dielectric materials, an applied electrical field will induce a local polarisation that varies across the channel due to steric constraints and over-screening effects [10, 210]. From the preceding discussion, this polarisation gradient is, in fact, reduced due to the presence of the diffusive process. The effect from the diffusion can be quantified from the dimensionless number $J = \tau_D v k_y^2$ [99]; using the values for water at ambient condition, the polarisation reduction is about 10% for wavelength of 3 nm [99].

Finally, it is worth mentioning that one can also define a transverse dipole moment autocorrelation function from the dipole x-component δp_x. It has been shown that the relaxation is significantly slower than the longitudinal relaxation at non-zero wavevector [100].

4.4.1 The Zero Wavevector Dielectric Permittivity

Recall that in the linear and static case the polarisation is given through the relation

$$\mathbf{P} = \rho \mathbf{p} = \varepsilon_0 \chi_e \mathbf{E}^{\text{ext}}. \tag{4.93}$$

As for the viscoelastic response, the system response to an electric field can be generalised with respect to time. To this end, we write the polarisation in terms of the electric susceptibility kernel,

$$\mathbf{P}(t) = \varepsilon_0 \chi_e \int_0^t \phi(t - t') \mathbf{E}^{\text{ext}}(t') \mathrm{d}t'. \tag{4.94}$$

Applying the convolution theorem, we get in frequency space

$$\widehat{\mathbf{P}}(\omega) = \varepsilon_0 \chi_e \widehat{\phi}(\omega) \widehat{\mathbf{E}}^{\text{ext}}(\omega). \tag{4.95}$$

The frequency-dependent susceptibility and the frequency-dependent dielectric permittivity are related through $\chi_e \widehat{\phi}(\omega) = \widehat{\varepsilon}_r(\omega) - 1$ [156], and the latter is given through the dipole moment autocorrelation function [60,32],

$$\widehat{\varepsilon}_r(\omega) = \frac{4\pi \rho_{\text{av}}^2}{k_B T \varepsilon_0} \left[C_{pp}(0,0) + i\omega \widehat{C}_{pp}(0,\omega) \right] + 1, \tag{4.96}$$

where, as usual, the Fourier–Laplace transformation of the correlation function is

$$\widehat{C}_{pp}(\mathbf{k}, \omega) = \mathcal{L}[C_{pp}(\mathbf{k}, t)], \tag{4.97}$$

for wavevector $\mathbf{k}$. The real part of the complex dielectric permittivity (the dielectric storage) is found by Fourier–Laplace transforming Eq. (4.91) and substituting, yielding the real part of the dielectric permittivity,

$$\varepsilon_r'(\omega) = \frac{4\pi \rho_{\text{av}}^2 C_{pp}(0,0)}{k_B T \varepsilon_0} \left(1 - \frac{\omega^2}{\omega_p^2 + \omega^2} \right) + 1, \tag{4.98}$$

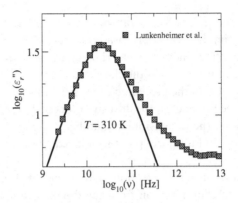

Figure 4.11 Dielectric loss for water at $T = 310\,\text{K}$ and ambient pressure. The x-axis is regular frequency (plotted on a logarithmic scale). Experimental data are taken from Ref. [2].

and the imaginary part (the dielectric loss),

$$\varepsilon_r''(\omega) = \frac{4\pi\rho_{\text{av}}^2 C_{pp}(0,0)}{k_B T \varepsilon_0} \frac{\omega_p \omega}{\omega_p^2 + \omega^2}, \tag{4.99}$$

such that $\widehat{\varepsilon}_r = \varepsilon_r' - i\varepsilon_r''$. This is on the exact same form as the Debye relaxation model originally derived on the basis of non-interacting microscopic dipoles with $\omega_p = 1/\tau_D$. For completeness, in the limit of zero frequency we recapture the well-known expression for the static dielectric permittivity,

$$\varepsilon_r = \frac{4\pi\rho_{\text{av}}^2 C_{pp}(0,0)}{k_B T \varepsilon_0} + 1. \tag{4.100}$$

In Fig. 4.11 the dielectric loss for water is plotted; both experimental values and predictions are shown for comparison. The theoretical prediction, Eq. (4.99), is given by the full line. It is seen that the theory does not correctly account for the high-frequency part of the spectrum (called the high-frequency wing). These frequencies are characteristic for molecular librational and bond-stretching modes and are often discussed on the basis of empirical functional forms.

4.4.2 The Static Dielectric Permittivity

We now turn to the limit of zero frequency and non-zero wavevector. In doing this, we will study the local field **E**, that is, the total electric field due to both screening and external field contributions. In this electrostatic limit, the curl of the electric field is zero:

$$\mathbf{\nabla} \times \mathbf{E} = \mathbf{0}. \tag{4.101}$$

As we only treat the case of zero free charge density, the divergence of the electric displacement field is zero, $\mathbf{\nabla} \cdot \mathbf{D} = 0$, and we have $\mathbf{E} = -\mathbf{P}/\varepsilon_0$. In Fourier space, Eq. (4.101) therefore reads

$$-\frac{i\mathbf{k}}{\varepsilon_0} \times \widetilde{\mathbf{P}} = \mathbf{0}, \tag{4.102}$$

which we for clarity will write out for $\mathbf{k} = (0, k_y, 0)$:

$$-\frac{i}{\varepsilon_0}\left(k_y\widetilde{P}_z, 0, k_y\widetilde{P}_x\right) = \mathbf{0}. \tag{4.103}$$

From this it is seen that in the static regime, the transverse components of the polarisation vanish for non-zero wavevector, and the static dielectric properties are therefore discussed in terms of the longitudinal dielectric permittivity.

It is indeed possible to theoretically discuss the polarisation relaxation in terms of the microscopic dipoles as we did previously. However, as pointed out by Bopp et al. [32], this is just the dipole term in the multipole expansion of the polarisation and fails at high wavevectors. The polarisation fluctuations are, in general, given by the bound charge density, ρ_b, Eq. (2.115):

$$\nabla \cdot \delta\mathbf{P} = -\delta\rho_b. \tag{4.104}$$

Bobb et al. [32] evaluated the bound charge density directly from the local charge density,

$$\delta\rho_b(\mathbf{r}, t) = \sum_i q_i\,\delta(\mathbf{r} - \mathbf{r}_i), \tag{4.105}$$

where q_i is the site charge. Since $\mathbf{k} = (0, k_y, 0)$, then in Fourier space

$$\widetilde{\delta P}_y = -\frac{\widetilde{\delta\rho_b}}{ik_y}, \tag{4.106}$$

for non-zero wavevector. The static polarisation autocorrelation function can then be defined in terms of the bound charge density,

$$S_{bb}(k_y) = \frac{1}{Vk_y^2}\left\langle \widetilde{\delta\rho_b}(k_y, 0)\widetilde{\delta\rho_b}(-k_y, 0)\right\rangle, \tag{4.107}$$

and can be thought of as a bound charge structure factor; hence, the symbol, S_{bb}.

It is possible to discuss the validity of the dipole moment expansion by comparing the bound charge structure factor with the longitudinal dipole moment autocorrelation function, Eq. (4.85). Figure 4.12(a) plots S_{bb} and $\rho_{av}^2 C_{pp}(k_y, 0)$ for the SPC/Fw water model, and it is clear that the two functions agree well for wavevectors below 2 Å^{-1}, where C_{pp} features a peak. For wavevectors larger than 2 Å^{-1} the agreement is very poor and functional forms are completely different.

The static (longitudinal) dielectric permittivity is given from the bound charge structure factor,

$$\frac{1}{\varepsilon_r(k_y)} = 1 - \frac{4\pi}{k_BT\varepsilon_0}S_{bb}(k_y). \tag{4.108}$$

Fig. 4.12(b) shows ε_r. As expected for zero wavevector, the permittivity is positive and we have $\varepsilon_r \approx 80$. However, for $k \approx 1$ Å^{-1} the permittivity becomes negative; this negative value is a fingerprint of an over-screening phenomenon at smaller length scales.

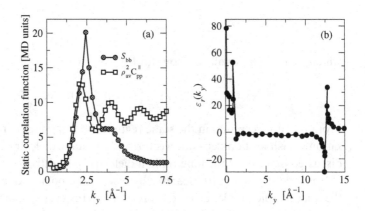

Figure 4.12 (a) Static correlation functions S_{bb} and $\rho_{av}^2 C_{pp}$ for the SPC/Fw water model at ambient conditions. (b) Kernel of the longitudinal dielectric permittivity for water (BJH model); data for (b) is taken from Ref. [32].

The bound charge structure factor is positive, $S_{bb} > 0$ for $k_y \neq 0$, which means according to Eq. (4.108) that

$$1 - \frac{1}{\varepsilon_r(k_y)} > 0. \tag{4.109}$$

This is fulfilled for

$$\varepsilon_r < 0 \text{ and } \varepsilon_r > 1, \tag{4.110}$$

defining a forbidden interval [58], which is evident in Fig. 4.12(b), where two singularities are observed at $k_y \approx 1 \text{ Å}^{-1}$ and $k_y \approx 12 \text{ Å}^{-1}$.

4.5 Further Explorations

1. In this exploration we will study liquid butane at state point $(\rho, T) = (582.3 \text{ kg m}^{-3}, 288 \text{ K})$. First, assume that the shear pressure relaxes in accordance with the Maxwell model, such that the shear pressure time correlation defined by

$$C(t) = \frac{1}{\tau_M} \int_0^\infty \overset{os}{P}_{yx}(t'+t) \overset{os}{P}_{yx}(t') \, dt'$$

is proportional to e^{-t/τ_M}.

Use Eq. (4.7), set the strain rate to zero, and apply initial condition $P_{yx}(0) = P_0$ to derive the exact result for C. Does this agree with simulation data? If not, consider how the Maxwell model can be extended to fit data better.

Computational resources available.

2. The viscosity kernel $\eta_0 \widetilde{f}(k_y)$ has not (yet) been studied for toluene. Use Eq. (4.48) to calculate $f(k_y)$ at state point $(\rho, T) = (879 \text{ kgm}^{-3}, 300 \text{ K})$. Estimate the minimum

length scale where $f \approx 1$. Is this length scale significantly different from water and butane, Fig. 4.5?

Discuss the functional forms, Eqs. (4.49) and (4.50). Which one fits toluene kernel data the best?

Computational resources available.

3. In this exploration we continue the investigation of the hydrodynamic correlation function $C_{\Omega\Omega}^{\|}$ for a generic diatomic molecular liquid. The atomic masses and bond length are unity (in molecular dynamics units).

    (i) Calculate the moment of inertia per unit mass, Θ, for the molecule.
    (ii) Run a simulation at state point $(\rho, T) = (0.85, 2.0)$. Evaluate $C_{\Omega\Omega}^{\|}$ as well as the (normal) pressure.
    (iii) Plot the corresponding dispersion relation, see Fig. (4.8), and from this estimate the transport coefficients η_r and ζ.
    (iv) Repeat (ii) and (iii) for different temperatures.

    How do η_r and ζ depend on temperature and pressure?

Computational resources available.

4. We continue the work with the generic diatomic system from Exploration 3. The molecule atoms now carry charges; one atom is positively charged and one negatively charged; hence, the molecules represent microscopic dipoles. We keep the system density and temperature fixed to $(\rho, T) = (0.85, 1.0)$.

    First, set the charges to unity in molecular dynamics units. Then, calculate the dipole moment autocorrelation function C_{pp} for different wavevectors. Plot the dispersion relation, Fig. 4.10, and from this extract the Debye relaxation time τ_D and the kinematic transport coefficient $v = v_t + v_l$. Repeat this for increasing charges, and explain your findings.

    Is v correlated with the single-particle diffusion coefficient?

Computational resources available.

5

Simple Nanoscale Flows

In Chapters 3 and 4 we discussed equilibrium systems and investigated the hydrodynamics in terms of equilibrium correlation functions. In this chapter we discuss standard flows like the planar Poiseuille and Couette flows, where the fluid is confined and driven by some external force. In nanoscale confinement a considerable fraction of the fluid molecules interact with the wall atoms, and this results in a fluid structuring in the wall–fluid interface. We must, of course, be concerned if the classical theory that assumes isotropy, homogeneity, and locality can be used in these cases of extreme confinement; we will focus on this potential complication throughout the chapter.

Before discussing confined systems, however, it is informative to take one step back and first investigate flows without the presence of walls. From this we will derive an important result for the generalised response for flows.

In Chapters 3 and 4 we derived the dynamical equations for the fluctuating parts of the hydrodynamic quantities. In this chapter we focus on the average quantities which, importantly, need not be constant in time and space. Recall that a hydrodynamic quantity A can be written as $A = \rho\phi$, where ρ is the mass density and ϕ is the associated field variable per unit mass. Up to first order in fluctuations we then obtain

$$\langle A \rangle = \langle (\rho_{av} - \delta\rho)(\phi_{av} - \delta\phi) \rangle = \langle \rho_{av}\phi_{av} \rangle = \rho_{av}\phi_{av}, \qquad (5.1)$$

since $\langle \delta\rho \rangle = \langle \delta\phi \rangle = 0$. The dynamics is again given by the balance equation, Eq. (2.2). Using the linear property of the ensemble average and ignoring higher-order fluctuations, we have

$$\frac{\partial}{\partial t}\rho_{av}\phi_{av} = \sigma_{av} - \boldsymbol{\nabla} \cdot (\rho_{av}\mathbf{u}_{av}\phi_{av}) - \boldsymbol{\nabla} \cdot \mathbf{J}_{av}. \qquad (5.2)$$

$\mathbf{J}_{av}$ is the average flux resulting from non-zero average driving forces present in the system. For the mass balance equation we have $\phi_{av} = 1$ and $\sigma_{av} = 0$, and we get

$$\frac{\partial \rho_{av}}{\partial t} = -\boldsymbol{\nabla} \cdot (\rho_{av}\mathbf{u}_{av}). \qquad (5.3)$$

For ease of reading, we will from now on omit the subscript "av" unless we actually study the fluctuations.

In Chapter 3 we argued that higher-order fluctuations in equilibrium can be ignored as long as we are not near the critical point, since the fluctuations are due to small thermal perturbations. While this also holds for the problems we treat in this chapter, the argument is, in fact, not general for non-equilibrium systems [45]; in Chapter 6 we will see one effect of fluctuations in a non-equilibrium system.

5.1 Homogeneous Flows

In this chapter we will see examples of flows generated by application of some external force acting in the x-direction. We study the resulting shear force or, equivalently, the shear pressure on fluid surfaces normal to the z-direction, that is, the only non-zero average shear component is $P_{xz} = P_{xz}(z)$. The force amplitude is sufficiently low; hence, the flows are, as usual, laminar, and the normal pressure gradient in the x-direction is negligible. In this first section we explore the fictitious situation where there are no confining walls, and we envision the system to be infinite in extent; see Fig. 5.1.

In Section 4.2 the non-local model for shear pressure was introduced. In the current geometry we have the generalised Newtonian law of viscosity,

$$P_{xz}(z) = -2\eta_0 \int_{-\infty}^{\infty} f(z - z') \dot{\gamma}(z') \, dz', \tag{5.4}$$

where the strain rate is

$$\dot{\gamma} = (\overset{os}{\nabla}\mathbf{u})_{xz} = \frac{1}{2} \frac{\partial u_x}{\partial z}. \tag{5.5}$$

In the following examples we see how different imposed strain rates affect the fluid shear pressure non-local response. The first two examples are from Ref. [198].

Example I Consider the situation where the strain rate is constant $\dot{\gamma}(z) = \dot{\gamma}_0/2$; this corresponds to a homogeneous Couette flow. The non-local response model yields

$$P_{xz}(z) = -\eta_0 \dot{\gamma}_0 \int_{-\infty}^{\infty} f(z - z') \, dz' = -\eta_0 \dot{\gamma}_0, \tag{5.6}$$

by virtue of Eq. (4.37). This is the same result as the local model, and we can immediately conclude that non-local viscous effects are not present for constant strain rates.

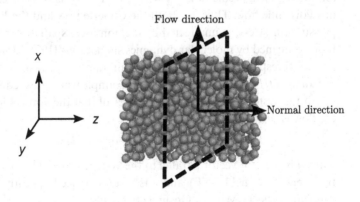

Figure 5.1   Schematic illustration of the virtual fluid surface. The arrows indicate the flow direction and the normal direction.

Example II Let us generalise Example I and let the strain rate be any power function of the form

$$\dot{\gamma}(z) = \frac{\dot{\gamma}_0}{2} z^n, \tag{5.7}$$

where n is a natural number. Applying the substitution $u = z - z'$, Eq. (5.4) is written as

$$P_{xz}(z) = -\eta_0 \dot{\gamma}_0 \int_{-\infty}^{\infty} f(u)(z-u)^n \, du. \tag{5.8}$$

To proceed, we expand the power $(z-u)^n$ as an alternating binomial series,

$$(z-u)^n = \sum_{k=0}^{n} (-1)^k \binom{n}{k} u^k z^{n-k}$$

$$= \left[\binom{n}{0} z^n - \binom{n}{1} u z^{n-1} + \binom{n}{2} u^2 z^{n-2} - \dots \right]. \tag{5.9}$$

Therefore,

$$P_{xz}(z) = -\eta_0 \dot{\gamma}_0 \left[z^n \int_{-\infty}^{\infty} f(u) \, du - \binom{n}{1} z^{n-1} \int_{-\infty}^{\infty} f(u) u \, du \right.$$

$$\left. + \binom{n}{2} z^{n-2} \int_{-\infty}^{\infty} f(u) u^2 \, du - \dots \right]. \tag{5.10}$$

Now, $\int_{-\infty}^{\infty} f(u) \, du = 1$ and f is even, thus all odd moments of f are zero. This leads to

$$P_{xz}(z) = -\eta_0 \dot{\gamma}_0 (z^n + a_2 z^{n-2} + a_4 z^{n-4} + \dots), \tag{5.11}$$

where the coefficients are given by the even moments of f,

$$a_k = \binom{n}{k} \int_{-\infty}^{\infty} f(u) u^k \, du \quad (k \text{ even}). \tag{5.12}$$

The first term in Eq. (5.11) is the local prediction, and the effect of the spatial correlations is then given by the even moments of the kernel, that is, by Eq. (5.12). This result can be generalised for any strain rate which is analytical, that is, where one can write $\dot{\gamma}(z) = \sum_n \dot{\gamma}_n z^n$.

For $n = 1$ the series Eq. (5.11) truncates after the first term, and the shear pressure follows the local predictions, $P_{xz} = -\eta_0 \alpha z$. This situation corresponds to a homogeneous Poiseuille flow. Thus, for both the Couette flow and the Poiseuille flow the shear pressure (or stress) is unaffected by the non-local spatial correlations, which has also been confirmed by molecular dynamics simulations [198]. More generally, an effect of non-local response requires that the strain rate profile features non-zero curvature.

Example III It is instructive to see an example where this requirement is fulfilled, and revisit the system from Section 1.1.5. Recall that the applied force density generating the flow is

$$\mathbf{F}^e(z) = \rho g_0 \cos(kz)\mathbf{i}, \tag{5.13}$$

where $\mathbf{i}$ is the unit vector parallel to the system x-axis. This force is called a sinusoidal transverse force field (STF) and was first introduced in synthetic molecular dynamics simulations (s-NEMD) by Gosling et al. [80]

Since we are in the laminar flow regime, the momentum balance equation reads

$$\rho \frac{\partial \mathbf{u}}{\partial t} = \mathbf{F}^e - \boldsymbol{\nabla} \cdot \mathbf{P}. \tag{5.14}$$

In the steady state, this reduces to

$$\boldsymbol{\nabla} \cdot \mathbf{P} = \rho g_0 \cos(kz)\mathbf{i}. \tag{5.15}$$

For our geometry and low Reynolds number Eq. (5.15) is a scalar equation,

$$\frac{\partial P_{xz}}{\partial z} = \rho g_0 \cos(kz). \tag{5.16}$$

Integration leads to the result in Eq. (1.14), but this time in terms of the shear pressure:

$$P_{xz}(z) = \frac{\rho g}{k} \sin(kz). \tag{5.17}$$

Notice that $\mathbf{P} = -\boldsymbol{\sigma}^T$. This result is derived from the momentum balance equation and does not involve any transport coefficient. We consider this exact.

We can compare the local and non-local descriptions. If only the fundamental mode in the velocity field is excited, that is, if

$$u_x(z) = \widetilde{u}_x \cos(kz), \tag{5.18}$$

where $\widetilde{u}_x$ is the fundamental Fourier coefficient, then the strain rate is

$$\dot{\gamma}(z) = -\frac{k\widetilde{u}_x}{2} \sin(kz). \tag{5.19}$$

The local description then follows immediately from Newton's law of viscosity, $P_{xz} = -2\eta_0 \dot{\gamma}$:

$$P_{xz}(z) = \eta_0 k \widetilde{u}_x \sin(kz). \tag{5.20}$$

In the non-local model, we will for simplicity let the kernel in real space be modelled by a single Gaussian function,

$$f(z) = \sqrt{\frac{\sigma}{\pi}} e^{-\sigma z^2}, \tag{5.21}$$

in which $1/\sqrt{\sigma}$ is a length scale and a measure of the kernel width. Substitution of Eqs. (5.21) and (5.19) into Eq. (5.4) gives

$$P_{xz}(z) = \eta_0 k \widetilde{u}_x e^{-k^2/4\sigma} \sin(kz). \tag{5.22}$$

Since $e^{-k^2/4\sigma} < 1$ for $k > 0$, the generalised response theory predicts a lower shear stress than the local theory for non-zero wavevectors. This agrees with findings from simulations, Fig. 1.7; for simple atomic liquids, σ is around $1/2$ Å^{-2} [102], and the shear pressure is reduced by half for wavelengths of around 0.5 nm, or a little more than one atomic diameter. In general, this reduction is significant if

$$\sigma > k^2, \tag{5.23}$$

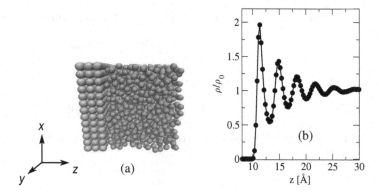

Figure 5.2 Molecular dynamics simulation of a confined liquid methane system. The bulk density is $\rho_0 = 445\,\mathrm{kg\,m^{-3}}$ and $T = 166\,\mathrm{K}$. (a) The simulated system setup. (b) The normalized density profile for the methane liquid.

that is, if the width of the kernel is large compared to the characteristic strain rate variation. For more general discussions where higher-order Fourier modes are excited, the reader is referred to Refs. [49, 200].

5.2 Effects of Confinement

Before studying flows, it is worthwhile to show some of the effects confinement has on both the fluid structure and the dynamics in equilibrium. It is not the purpose here to reiterate the statistical mechanical theories that exist on the topic or to give an exhausting review, but to show the examples relevant for our later treatment.

5.2.1 Fluid Structuring

Figure 5.2(a) shows a direct molecular dynamics simulation (d-NEMD) of liquid methane next to a wall. The wall is composed of spherical atoms that are arranged on a simple, or primitive, cubic lattice, and the coordinate system is chosen such that the centre of mass of the left-most wall layer is positioned at $z = 0$. We return to the choice of coordinate system later. The wall atoms are not fixed, but tethered to crystal sites by a restoring spring force, and during the simulation the wall atoms are thermostated, keeping the system at a constant average temperature.

Often, in simulations the steady-state quantities are calculated from the corresponding time averages; that is, for the density we can formally write this in terms of the microscopic definition as

$$\rho(z) = \frac{1}{\tau_{\mathrm{obs}}} \int_0^{\tau_{\mathrm{obs}}} \rho(z,t)\,\mathrm{d}t = \frac{1}{A}\left\langle \sum_i m_i \delta(z - z_i)\right\rangle_t, \tag{5.24}$$

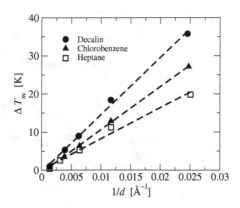

Figure 5.3   Melting temperature depression as function of pore diameter. Re-plotted from Jackson and McKenna [118].

where τ_{obs} is the simulation or observation time, A is the (x,y)-plane area of the simulation box, and the brackets represent the time average as indicated by the subscript t. In practice, the Dirac delta is replaced with a so-called bin method, where the system is divided into a number of bins and the time average mass density in the bin is then calculated. The density profile is shown in Fig. 5.2(b). The first striking effect of the confinement is the induced density variation adjacent to the wall [117, 201]. The molecular diameter of methane is approximately 3.7 Å. This is also the characteristic length scale for the density variations, indicating that the variation is related to molecular layer structuring as also reported by Horn and Isrealachvili [114].

This inherent layering is also observed in the radial distribution function g for non-confined fluid (see Fig. 3.10); hence, the fluid molecules tend to pack against the wall atoms in a manner similar to the bulk fluid. In fact, the statistical mechanical theories [90, 156] for the fluid concentration profile is based on a perturbation of the radial distribution function. For later use we will discuss the layering in more simple terms, namely through the Boltzmann potential function, φ_c, such that the density is given by

$$\rho(z) = \rho_0 e^{-\varphi_c/k_B T}. \tag{5.25}$$

The subscript c indicates that this potential function is associated with confinement. The potential function depends on the wall density, the specific wall–fluid interactions, and the wall–fluid geometrical commensurability [117, 165]. It must fulfil

$$\varphi_c \to \infty \text{ as } z \to 0 \text{ and } \varphi_c \to 0 \text{ as } z \to \infty \tag{5.26}$$

and can be evaluated once the density profile is known or by theoretical means, as we see in Chapter 6 for electrolytes.

Depending on the specific details of the system, the wall may introduce an interfacial energy that prevents the liquid from crystallising. This is known as the Gibbs–Thomson effect and can be measured experimentally from the melting temperature depression ΔT_m defined as the difference between the bulk melting temperature and the measured melting temperature in confinement. Figure 5.3 shows data for the melting temperature depression for small-weight organic fluids [118] in porous glass with diameters in the

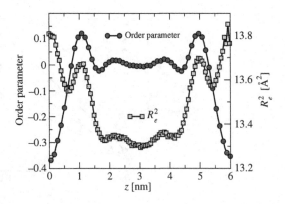

Figure 5.4 Order parameter and the squared molecular end-to-end distance for confined butane. From Ref. [102] with permission.

range 4–73 nm. The inverse relation between the depression temperature and the pore diameter d is in agreement with the theoretical prediction $\Delta T_m \propto 1/d$.

Importantly, a melting temperature increase can also be observed, $\Delta T_m < 0$, in systems where the wall–fluid interaction is energetically favourable compared to the fluid–fluid interactions (e.g. for very dense walls). The enhanced adhesion gives rise to crystallisation in the wall–fluid interfacial region even above the melting temperature; see Ref. [68] for examples of this case.

5.2.2 Molecular Alignment

For molecular fluids the presence of a wall will induce molecular alignment. This alignment can be quantified by the local order parameter; see Ref. [78]:

$$S = \frac{1}{2}\left(3\left\langle \cos^2(\theta)\right\rangle - 1\right), \qquad (5.27)$$

where θ is the angle between a specific molecular vector and the wall surface plane. For water, a natural choice of such a molecular vector is the dipole moment vector, and for butane it can be the end-to-end vector denoted R_e. If $S = -1/2$, the molecular vector is parallel to the wall, for $S = 1$, it is normal to the wall, and for $S = 0$, the molecular vector is randomly distributed and, hence, the molecules are randomly oriented.

Figure 5.4 shows molecular simulation data for the order parameter profile for liquid butane confined in a 6 nm slit-pore. The order parameter is defined from the end-to-end vector. It is seen that in the wall–fluid interface the butane molecules are oriented and on average align with the wall surface plane in the first 5–8 Å. After this layer, the molecules then organize normal to the wall in the next layer, before becoming randomly distributed.

The figure also shows the local average end-to-end vector, that is, a measure of the average butane molecular length. It is seen that the molecules stretch as they are close to the wall; this is likely due to the constrained intra-molecular angle rotation in the region. Alignment phenomena are also reported for other molecular fluids; for water, see, for example, Refs. [158, 210].

5.2.3 Dynamics

If we return to the density profile in Fig. 5.2(b) we expect the transport coefficients to vary significantly in the wall–fluid region, as they are state point dependent. For example, the self-diffusion coefficient introduced in Chapter 3 depends on the density, implying that as we consider the diffusion in the confined direction, we have $D_s = D_s(z)$. Moreover, according to Fick's law we associate molecular diffusion with a directional motion down the concentration gradient; and this motion, we can expect, is hindered in some way by the presence of the wall. That is, in general, the diffusion coefficient in the confined direction, $D_s^{\perp}$, is different from the coefficient in the parallel direction, $D_s^{\parallel}$. This anisotropy is equivalent to the dielectric permittivity discussed in Chapter 1, and the diffusion should be described in terms of a rank-2 diffusion tensor. Therefore, in the current slit-pore geometry, Fick's law for the single-particle flux is written as

$$\mathbf{J}^i = -\mathbf{D}_s(z) \cdot \nabla \rho_i, \tag{5.28}$$

where $\mathbf{D}_s$ is the self-diffusion tensor,

$$\mathbf{D}_s(z) = \begin{bmatrix} D_s^{\parallel}(z) & 0 & 0 \\ 0 & D_s^{\parallel}(z) & 0 \\ 0 & 0 & D_s^{\perp}(z) \end{bmatrix}, \tag{5.29}$$

if we assume that the flux in one direction is only dependent on the gradient in that particular direction, that is, we ignore any cross-coupling effects.

Magda et al. [151] studied the diffusion of a simple Lennard–Jones fluid confined in nano-slit-pores of different heights. The wall atoms are not explicitly modelled, and the wall–fluid interactions are given by a smooth LJ-type potential. In the case where h is approximately 11 particle diameters, the authors calculated the parallel diffusion coefficient $D_s^{\parallel}$ in different slabs in the fluid; in Fig. 5.5 these are indicated as regions I, II, and III. It was found that the diffusion coefficient did not vary significantly with respect to the slab, and is the same as the bulk value D_s. $h = 11$ corresponds to around 11 molecular diameters, or approximately 4.1 nm in the case of methane. Even in the very-high-density region in the wall–fluid interface the particles possess the same mobility as in the bulk. Thus, for this system the diffusion tensor parallel components are constants, and the single-particle diffusion dynamics parallel to the confinement direction are given through the usual scalar expression discussed in Section 3.2.

An important note: for constant chemical potential, Magda et al. [151] also showed that the average (or effective) diffusion coefficient is reduced below $h = 11$.

To account for this surprisingly small effect of the local density on the local transport properties, Bitsanis et al. proposed a local average density model (LADM) [21]. The fundamental assumption is that the transport coefficients are functions of the spatial averaged density, which we denote $\overline{\rho}$, and not the local density at the point. For the confined situation above the local average density is computed from the convolution

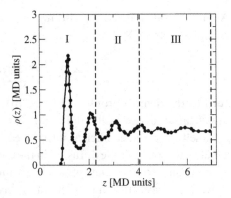

Figure 5.5 Density profile for a Lennard–Jones fluid confined in a slit-pore with smooth walls. Only part of the fluid density profile near one wall is shown. The Roman numerals indicate slab regions where the diffusion coefficient parallel to the wall has been evaluated. For I: $D_s^{||} = 0.114 \pm 0.004$. II: $D_s^{||} = 0.109$ III: $D_s^{||} = 0.111$; units are in reduced molecular dynamics units. Data are replotted from Ref. [151].

$$\bar{\rho}(z) = \frac{1}{\sigma} \int_{-\sigma/2}^{\sigma/2} w(z - z')\rho(z')\,dz' , \qquad (5.30)$$

where σ is one molecular diameter and w is a weighting dimensionless kernel. Bitsanis et al. [22] suggested the kernel

$$w(z) = 3/2 - 6(z/\sigma)^2 , \qquad (5.31)$$

but many other functional forms have been used; see, for example, Ref. [110]. Figure 5.6 shows the density ρ and local average density $\bar{\rho}$ as functions of z-coordinate; $\bar{\rho}$ is calculated from the weighting function given in Eq. (5.31). In this way the density variations are suppressed considerably, and using the LADM to predict, for example, the single-particle mass diffusion and the local stress agrees well with the results from direct-NEMD simulations [22, 110]. While the LADM has been applied successfully to various problems, it cannot account for the reduced stress in STF systems that we discussed in the previous section because the density is constant in this case. One may therefore conjecture that the reduced effect of the large density variation is due to the non-local nature of the transport properties leading to the generalized hydrodynamics formalism discussed in Section 4.2 rather than a non-local density effect. Only a few modelling attempts using generalised hydrodynamics have been made for confined fluids [41, 216, 217], and this remains an open question.

For more complex fluids and even smaller confinements, there is a change in dynamics. Milischuk et al. used molecular dynamics to study water confined in silica nanotubes [157, 158]. They found that the hydrogen bonding network between water molecules is significantly affected by the silica tube, and that the diffusion coefficient parallel to the tube's longitudinal axis is reduced as a function of tube radius: for a tube radius of 1 nm, $D_s^{||} = 1.6 \times 10^{-9}$ m^{2}s^{-1}, and for a radius of 2 nm, $D_s^{||} = 2.1 \times 10^{-9}$ m^{2}s^{-1}. Importantly, the local diffusion coefficient at the wall–fluid interface is very

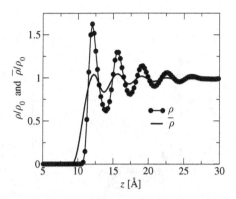

Figure 5.6 Molecular dynamics for methane. Comparison between the local density profile and the local average density profile.

low compared to the diffusion in the tube mid-point, likely because of the different local hydrogen bond network.

Diffusion pertains to the molecular translational motion. The molecular rotational dynamics, or specifically, the orientational relaxation has also been studied. Rather than using the local order parameter defined by de Gennes and Prost, Eq. (5.27), one can also investigate the relaxation from [90],

$$C_L(t) = \langle P_L \left(\hat{\mathbf{u}}(t) \cdot \hat{\mathbf{u}}(0) \right) \rangle, \tag{5.32}$$

where P_L is the Legendre polynomial of degree L, and $\hat{\mathbf{u}}$ is an intra-molecular unit vector which must be specified. C_2 can be inferred from IR spectroscopy and a neutron scattering experiment. Again, Milischuck and Ladanyi [158] used the OH-bond vector and the second-order Legendre polynomial $P_2(x) = (3x^2 - 1)/2$, showing that C_2 is greatly affected by the confinement slowing down the orientational relaxation, that is, the single molecular orientation becomes less randomized with respect to time as the confinement increases.

Finally, we compare the collective density fluctuations for a confined system with the corresponding non-confined system. The system is in equilibrium, and we can use the theory developed in Chapter 3 to calculate density autocorrelations along a wavevector parallel to one of the non-confined directions in the slit-pore. In Fig. 5.7 the dynamic structure factor is plotted for liquid methane for a single wavevector parallel to the wall, $\mathbf{k} = (0, k_y, 0)$. The slit-pore height is $h = 3.26$ nm. For the comparison, the dynamic structure factor for a bulk non-confined system is also plotted. It can be seen that the density correlations are significantly changed. From the classical hydrodynamic theory developed in Chapter 3 we can conclude the following: First, the Brillouin peak is shifted to lower frequencies in the confined situation, Fig. 5.7(a). This means that the adiabatic speed of sound is reduced. By normalisation of the curves with respect to the Rayleigh peak height, Fig. 5.7(b), we observe a very small increase in the Rayleigh peak half-width, $\Delta\omega_{\mathrm{Ra}}$; hence, the thermal diffusivity D_T is slightly increased in the confined case. Finally, shifting the frequency and normalising with respect to the Brillouin peak

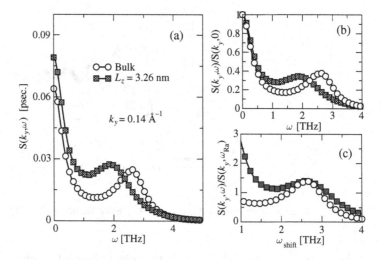

Figure 5.7 Dynamic structure factor obtained from molecular dynamics for model liquid methane at $\rho = 445\,\mathrm{kg\,m^{-3}}$ and $T = 166\,\mathrm{K}$. Both the confined and non-confined (bulk) situations are plotted. The wavevector is parallel to the unconfined y-direction.

height, Fig. 5.7(c), reveal that the attenuation coefficient Γ also increases slightly for confinement, that is, the dampening of the density wave is increased.

In summary: The presence of a wall introduces a wall–fluid interface (or region) wherein the fluid density varies significantly and where molecular alignment effects are present. The effects of this interface are manifold and the effect magnitudes depend on the system details, such as wall density, wall–fluid interactions, the wall crystal structure, and so forth. Importantly, the complications in the wall-fluid interface are not included in the classical hydrodynamic picture, and we must proceed with our (hydrodynamic) exploration keeping the findings of this section in mind.

5.3 The Planar Poiseuille Flow and the Navier Boundary Condition

One of the most well studied nanoscale fluid flow is the Poiseuille flow [136, 205, 206]. In the *planar* Poiseuille flow, the fluid is confined between two planar walls, thus resembling the slit-pore geometry, and the flow is driven by an external force (especially relevant in simulations) and/or a constant non-zero pressure gradient. In either case there will be a net force density acting on the fluid, and we will, as usual, denote this by $\rho\mathbf{g}$. We again choose a coordinate system such that the planar walls lie in the (x,y)-plane and the direction of confinement is in the z-direction; see Fig. 5.8(a). We will here assume that the antisymmetric shear pressure can be ignored, and focus on the laminar case where both the Mach and Reynolds numbers are sufficiently low. Also, any viscous heating is conducted away from the system through the wall; hence, the temperature

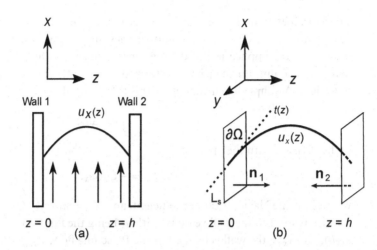

Figure 5.8 (a) Schematic illustration of the planar Poiseuille flow driven by a force density field (arrows) acting in the x-direction. Note, the y-axis is normal to the paper. (b) Geometrical illustration of the Navier boundary condition. See also Fig. 1.1.

is constant. Then from the momentum balance equation and Newtonian viscosity law we have the Navier–Stokes equation for the streaming velocity x-component (recall, strictly the ensemble averaged streaming velocity),

$$\rho \frac{\partial u_x}{\partial t} = \rho g + \eta_0 \frac{\partial^2 u_x}{\partial z^2}. \tag{5.33}$$

In the steady state, $u_x(z,t) = u_x(z)$ and we arrive at the Stokes equation,

$$\frac{\mathrm{d}^2 u_x}{\mathrm{d}z^2} = -\rho g / \eta_0. \tag{5.34}$$

To solve this problem, we need to specify the boundary conditions. As shown in Chapter 1, the velocity slip at the wall becomes important for nanoscale flows, and we cannot, in general, assume no-slip boundaries $u_x(0) = u_x(h) = 0$. As a first approach, we can apply a more general Dirichlet boundary condition, that is,

$$u_x(0) = u_x(h) = u_w. \tag{5.35}$$

The no-slip boundary condition is, of course, just a special case of the Dirichlet boundary where $u_w = 0$. Then, integrating Eq. (5.34), we arrive at the general solution,

$$u_x(z) = -\frac{\rho g}{2\eta_0} z^2 + K_1 z + K_2, \tag{5.36}$$

where K_1 and K_2 are constants of integration. Application of the boundary conditions yields the particular solution:

$$u_x(z) = \frac{\rho g}{2\eta_0} z(h - z) + u_w. \tag{5.37}$$

The Dirichlet boundary condition is ad hoc in the sense that u_w must be known a priori; for example, from molecular dynamics simulations. A more appealing approach, perhaps, is the application of the Navier boundary condition. The fundamental idea here is to account for the frictional force the wall with area A exerts on the fluid at the boundary. A simple linear model for this is [143, 163]

$$F_w = -A\zeta_N \Delta u_x, \tag{5.38}$$

or, by division with the area,

$$P_{xz} = -\zeta_N \Delta u_x, \tag{5.39}$$

where ζ_N is the Navier friction coefficient, and $\Delta u_x = u_x - V_{\text{wall}}$ is the relative fluid velocity with respect to the wall velocity, with u_x being the fluid velocity at the wall, $u_x(0)$ or $u_x(h)$, and V_{wall} the wall velocity. For the Poiseuille flow, $V_{\text{wall}} = 0$. The pressure tensor index xz indicates that we study a shear pressure at the boundary surface with normal vector parallel to the z-axis and the force x-component.

Due to continuation of the shear pressure, we also have that $P_{xz} = -\eta_0 \partial u_x / \partial z$ at the wall, and hence for the wall at, say, $z = 0$ we have

$$u_x(0) = \frac{\eta_0}{\zeta_N} \left. \frac{\partial u_x}{\partial z} \right|_{z=0}. \tag{5.40}$$

The fraction on the right-hand side defines a characteristic length scale which is denoted the slip length, L_s,

$$L_s = \eta_0 / \zeta_N. \tag{5.41}$$

We can generalize the Navier boundary condition. If $\partial\Omega$ is the boundary surface and $\mathbf{n}$ the normal to $\partial\Omega$ pointing into the fluid, see Fig. 5.8 (b), then

$$L_s(\boldsymbol{\nabla} u_x) \cdot \mathbf{n} = \Delta u_x, \tag{5.42}$$

for point $(x, y, z) \in \partial\Omega$. From this we see that the Navier boundary condition is simply a Neumann boundary condition.

Note: For the Poisuille flow, the slip length can be interpreted as a distance away from the wall–fluid boundary where the linearly extrapolated velocity is zero. The extrapolation is done via the tangent line $t = t(z)$ at the boundary; see Fig. 5.8(b).

To find the solution to this boundary value problem, we must then solve the differential equation Eq. (5.34) with Neumann boundary conditions, Eq. (5.42),

$$L_s \left. \frac{\partial u_x}{\partial z} \right|_{z=0} = u_x(0) \quad \text{and} \quad L_s \left. \frac{\partial u_x}{\partial z} \right|_{z=h} = -u_x(h). \tag{5.43}$$

The general solution is given in Eq. (5.36), and we arrive at the particular solution in terms of the slip length,

$$u_x(z) = \frac{\rho g}{2\eta_0} \left(h(z + L_s) - z^2 \right). \tag{5.44}$$

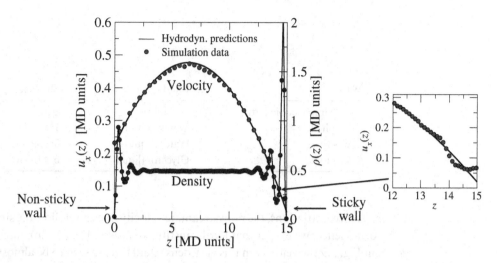

Comparison between the predicted velocity profile and d-NEMD simulation data for a planar Poiseuille flow system. For the hydrodynamic predictions, all parameters are found independently of the NEMD simulation, and no additional fitting is carried out. From Ref. [105] with permission.

In general, the two walls need not be identical and can be characterized by two different slip lengths, $L_s^{(1)}$ and $L_s^{(2)}$ for wall 1 and wall 2, respectively. In this case the solution is shown to give

$$u_x(z) = \frac{\rho g}{2\eta_0}\left(B(z + L_s^{(1)}) - z^2\right), \qquad (5.45)$$

where

$$B = \frac{h(h + 2L_s^{(2)})}{L_s^{(1)} + L_s^{(2)} + h}. \qquad (5.46)$$

Hansen et al. [105] performed non-equilibrium simulations of a Lennard–Jones fluid undergoing a planar Poiseuille flow. The wall–fluid interactions at wall 1 were the same as the fluid–fluid interactions, whereas the wall–fluid interactions at wall 2 had an additional attractive part. In this way the energetic interaction is controlled and leads to a more 'sticky' surface at wall 2. The two corresponding slip lengths and the fluid viscosity were found from independent methods, enabling a direct comparison with the non-equilibrium simulation without any free fitting parameters; see Fig. 5.9. The streaming velocity in molecular dynamics simulations is calculated from the definition

$$\mathbf{u}(z) = \frac{\langle \sum_i m_i \mathbf{v}_i \delta(z - z_i)\rangle_t}{\langle \sum_i m_i \delta(z - z_i)\rangle_t}, \qquad (5.47)$$

where i is the fluid particle index, and the brackets represent a time average. As we have mentioned, the Dirac delta is, in practise, often replaced by a bin method.

We see that the molecular dynamics results and hydrodynamic prediction agree quite well in the channel interior. Note, however, that at wall 2, where the density features

Table 5.1 Slip lengths for four different systems. σ is the Lennard–Jones length scale. The reader is referred to the references for further information. [a] Ref. [105], [b] Ref. [125] (also see refs. therein), [c] Ref. [135].

Wall	Fluid	Slip length, L_s
[a] LJ - simple cubic lattice	LJ (fluid)	$2\,\sigma$
[b] Graphene	Water (liquid)	1–80 nm
[c] Au - FCC	Water (liquid)	36–39 nm
[c] Au - FCC	Glycerol (liquid)	6×10^3 nm

extremely large variations, the fluid streaming velocity does not follow the simple quadratic form; hence, we expect that local constitutive model, Eq. (3.22b), fails. From the zoom in Fig. 5.9 the velocity in this very dense fluid layer is seen to be almost constant, that is, the fluid motion is like a 'solid sliding brick'. Of course, in this extreme case where the wall–fluid region behaves solid-like, we must expect that the classical theory performs poorly. Moreover, since the strain rate is non-linear in this region, we also expect that non-local response phenomena become important in accordance with our discussion in Section 5.1.

Another important point which is repeated here is that the hydrodynamics model consists of a dynamical equation *and* a set of appropriate initial/boundary conditions. Having a slip boundary condition is not an indication of a breakdown of hydrodynamics. For use later, Table 5.1 lists a few values for the slip length for different wall–fluid systems. The slip length spans many orders of magnitude and is a function of the wall–fluid interactions and geometrical commensurability [179, 193].

Thompson and Troian [194] performed a series of d-NEMD simulations of a Lennard–Jones system undergoing a Couette flow; we will treat this flow in the following section. In their simulations they changed the strain rate and found that the slip length followed the empirical law

$$L_s = L_s^0 (1 - \dot{\gamma}/\dot{\gamma}_c)^{-1/2}, \tag{5.48}$$

where $2\dot{\gamma} = \partial u_x / \partial z$ is the strain rate at the wall-fluid boundary, L_s^0 is the slip-length in the limit of zero strain-rate, and $\dot{\gamma}_c$ is the so-called critical strain rate, such that when the strain rate approaches $\dot{\gamma}_c$, the slip length diverges and the Navier boundary model breaks down. It is interesting that for the Lennard–Jones model the non-linear dependency of the strain rate happens at very large strain rates,[1] yet at strain rates where the viscosity is constant; that is, the fluid features Newtonian behavior, but the slip length is in the non-linear regime. Priezjev [177] also reported a simple correlation between the slip length and the inverse of the in-plane structure factor, which again highlights the fact that the geometrical commensurability greatly affects the slip.

One obvious effect of the slip is flow enhancement, which is quantified by the enhancement coefficient, E^{slip}, already introduced in Eq. (1.6). First, we need to find

---

[1] Which cannot be achieved in the lab

the volumetric flow rate, Q, which is the fluid volume crossing a surface normal to the flow direction per unit time; in our geometry, the surface lying in the yz-plane, Fig. 5.8. Let the surface have area A; then

$$Q = \iint_A u_x(y,z)\,\mathrm{d}y\mathrm{d}z. \tag{5.49}$$

For a planar Poiseuille flow and if walls 1 and 2 are identical, we simply need to integrate Eq. (5.44):

$$
\begin{aligned}
Q^{\mathrm{slip}} &= \frac{\rho g}{2\eta_0} \int_0^w \mathrm{d}y \int_0^h h(L_s+z) - z^2 \,\mathrm{d}z \\
&= \frac{\rho g w}{12\eta_0}(h^3 + 6h^2 L_s),
\end{aligned}
\tag{5.50}
$$

where w is the system length in the y-direction. The flow enhancement due to the slip is then simply

$$E^{\mathrm{slip}} = \frac{Q^{\mathrm{slip}}}{Q^{\mathrm{noslip}}} = 1 + 6L_s/h, \tag{5.51}$$

as was stated in Chapter 1.

If one considers water confined between two gold FCC sheets separated by 10 nm, *and* assumes that hydrodynamics is a good model for the flow, then the flow enhancement is around 24, using the slip length given in Ref. [135], Table 5.1. For slip lengths on the order of a few nanometres, which is a typical order of magnitude, the last term in Eq. (5.51) can be safely ignored even for micro-fluidic systems, and no enhancement is observed.

5.3.1 The Hydrodynamic Channel Height

One very important question has been postponed until now, namely "What exactly is the channel height h?" We must address this, as the validity of the hydrodynamic model depends on the choice of h, since this enters the solution to the Navier–Stokes equation. In Fig. 5.10 the fluid density in the wall–fluid interface is illustrated. As the minimum centre-of-mass distance between the wall and fluid particles is around one molecular diameter, σ, there exists a depleted region between the wall and the fluid. Two heuristic definitions of h that come to mind are (i) the distance between the centre of mass of the first wall layer in the two walls and (ii) the length over which the density profile is non-zero. These definitions will give different velocity profiles in nanoscale geometries.

A simple theoretical treatment of the question is given by Herrero et al. [108]. Here we focus on the case where the two walls are identical, implying $L_s^{(1)} = L_s^{(2)}$. As the fluid moves relative to the surface, we can calculate the wall shearing force, F_w, acting on the fluid from wall 2, where $z = h$; see Fig. 5.8(a). This is given by Newton's law of viscosity,

$$F_w = A\eta_0 \left.\frac{\partial u_x}{\partial z}\right|_{z=h}. \tag{5.52}$$

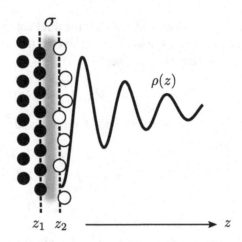

Figure 5.10 Illustration of the depleted region $z_1 < z < z_2$ (shaded area) between the wall and the fluid.

Again, A is the wall surface area. By differentiation of either Eq. (5.37) or Eq. (5.44), thus, approximating the fluid velocity to be in accordance with classical hydrodynamics, we then get h in terms of the shear force

$$h = -\frac{2F_w}{\rho g A}. \tag{5.53}$$

For steady flows, we can apply force balance, and h can be expressed on a very simple form. If the flow is driven by application of an external force, F_i^{ext}, acting on the fluid particles, the hydrodynamical force density is $\rho g = n F_i^{\text{ext}}$, where n is the system number density. The total applied force is then

$$F_{\text{tot}}^{\text{ext}} = \sum_i F_i^{\text{ext}} = \frac{\rho g N}{n}, \tag{5.54}$$

where N is the number of fluid molecules. In the steady state, $F_{\text{tot}}^{\text{ext}} + 2F_w = 0$, which is rearranged to give

$$h = \frac{N}{nA} \tag{5.55}$$

through Eq. (5.53). Interestingly, experience shows that the resulting h is approximately the heuristic definition (ii) if the fluid structure in the wall–fluid interface is not too strong. Definition (ii) is used in the hydrodynamic predictions shown in Fig. 5.9

5.3.2 Failure of the Classical Theory

Recall, Fig. 5.9 shows the velocity profile for a planar Poiseuille flow where the channel height is around 15 molecular diameters, that is, h is around 4–5 nm. It was concluded that classical hydrodynamical theory predicts the flow profile satisfactory for this situation. Travis et al. [205] studied the same system, but in a slit-pore with a height of just four diameters or just above one nanometre; Fig. 5.11(a) plots the streaming velocity

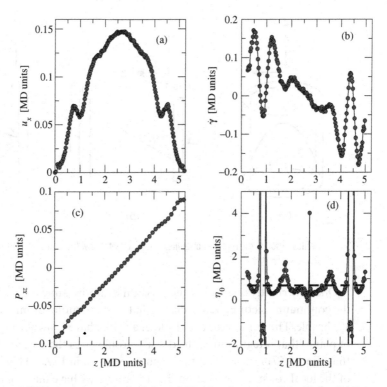

Figure 5.11 Results from a d-NEMD simulation of a Lennard–Jones fluid undergoing a planar Poiseuille flow. (a) The streaming velocity. (b) The strain rate. (c) Shear pressure. (d) Local viscosity has predicted by Newton's law of viscosity. The horizontal punctured line indicates the bulk value.

profile in this situation. One clearly observes that the profile features superimposed modulations and the classical theory is no longer a satisfactory model.

In fact, Travis et al. showed that under these extreme confinements Newton's law of viscosity breaks down. In the current geometry, this reads

$$P_{xz} = -2\eta_0 \dot{\gamma}, \tag{5.56}$$

where the strain rate is $2\dot{\gamma} = \partial u_x / \partial z$. $\dot{\gamma}$ can be calculated directly by numerical differentiation of the velocity profile; see Fig. 5.11(b). The shear pressure can be calculated from the momentum balance equation, which for the steady state reads

$$\frac{\partial P_{xz}}{\partial z} = \rho g. \tag{5.57}$$

Hence, from numerical integration of the mass density profile, P_{xz} is found; Fig. 5.11(c). Rearranging Eq. (5.56), we get for the local shear viscosity

$$\eta_0(z) = -P_{xz}(z)/\dot{\gamma}(z). \tag{5.58}$$

This is plotted in Fig. 5.11(d), showing singularities where the strain rate crosses the x-axis.

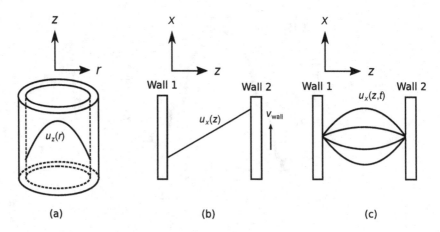

Figure 5.12   Schematic illustration of the Hagen–Poiseuille flow (a), Couette flow (b), and oscillatory flows (c).

One may rightly question if the classical hydrodynamic theory breaks down because the continuum picture breaks down, that is, if the momentum balance equation is not applicable. This is not the case; Todd et al. [197] have shown from the so-called method of planes that integration of the density profile gives the correct shear pressure profile; thus, Eq. (5.57) is valid, and the singularities seen in Fig. 5.11(d) are due to the failure of the local constitutive relation, Eq. (5.56). As we have mentioned, the non-local constitutive model for the shear pressure has been applied to confinement [41, 216, 217]; however, a general solution to the problem is still not available.

5.4 Other Simple Flows

In this section we will discuss other simple flows, specifically, (i) the Hagen–Poiseuille flow, (ii) the Couette flow, and (iii) an oscillatory flow. Figure 5.12 illustrates the system geometries. We shall assume that the classical theory is applicable. The flows are treated in most hydrodynamic text books; however, it is worth discussing them in the context of nano-confinement, as some points are important to highlight.

5.4.1 The Hagen–Poiseuille Flow

In the Hagen–Poiseuille flow the fluid flows in a cylindrical geometry, Fig. 5.12 (a), and is driven by a force density, ρg, that acts parallel to the longitudinal axis, z. Again, the force density can be due to an external force, a constant pressure gradient, or both. For low Reynold numbers only the velocity component u_z is non-zero, and this only depends on the radial coordinate r, hence, $u_z = u_z(r)$. The Navier–Stokes equation in cylindrical coordinates reads

$$\eta_0 \left(\frac{d^2 u_z(r)}{dr^2} + \frac{1}{r} \frac{du_z(r)}{dr} \right) = -\rho g. \tag{5.59}$$

From the chain rule, the left-hand side is re-written as

$$\frac{d^2 u_z(r)}{dr^2} + \frac{1}{r} \frac{du_z(r)}{dr} = \frac{1}{r} \frac{d}{dr} \left(r \frac{du}{dr} \right) \tag{5.60}$$

and we obtain

$$\frac{d}{dr} \left(r \frac{du}{dr} \right) = -\frac{\rho g}{\eta_0} r. \tag{5.61}$$

Integrating twice, we arrive at the general solution

$$u_z(r) = -\frac{\rho g}{4\eta_0} r^2 + K_1 \ln(r) + K_2, \tag{5.62}$$

where K_1 and K_2 are constants of integration. Since u_z cannot diverge for $r = 0$, we must require that $K_1 = 0$, and therefore

$$u_z(r) = -\frac{\rho g}{4\eta_0} r^2 + K_2 \tag{5.63}$$

with the Navier boundary condition

$$L_s \frac{du_z(r)}{dr} \bigg|_{r=R} = -u_z(R), \tag{5.64}$$

with R being the tube radius. The particular solution to this problem is

$$u_z(r) = \frac{\rho g}{4\eta_0} \left((R + 2L_s)R - r^2 \right). \tag{5.65}$$

The volumetric flow rate follows from this solution. We express the velocity dependency of the azimuthal angle explicitly and get

$$Q = \int_0^{2\pi} d\theta \int_0^R r u_z(r)\, dr$$
$$= \frac{\pi \rho g}{8\eta_0} (R^4 + 4L_s R^3). \tag{5.66}$$

This gives the slip enhancement coefficient

$$E^{\text{slip}} = 1 + 4L_s/R. \tag{5.67}$$

Things are more complicated; L_s depends on the tube radius, that is, $L_s = L_s(R)$. Figure 5.13 plots the slip length for the carbon nanotube and water system as a function of R. In the small radii regime, the slip length increases rapidly for decreasing R, that is, the flow enhancement grows much faster than R^{-1} from Eq. (5.67). Again, this is still an active research area.

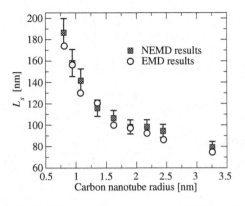

Figure 5.13 Slip length dependency of tube radius of a Hagen–Poiseuille flow of water in carbon nanotubes (obtained from molecular dynamics simulation). The slip length is found from methods based on both direct-NEMD and EMD simulations. Data are re-plotted from Ref. [125].

5.4.2 The Couette Flow

For the Couette flow, Fig. 5.12(b), the two planar walls move relative to one another. Without loss of generality, we let wall 1 be at rest and let wall 2 move in the x-direction with velocity V_{wall}. We have no pressure gradient in the x-direction or external forces acting, hence, $\rho \mathbf{g} = \mathbf{0}$. As usual, we assume that the Reynolds number is sufficiently low, hence the Navier–Stokes equation reduces to a Laplace equation,

$$\frac{\text{d}^2 u_x}{\text{d}z^2} = 0, \tag{5.68}$$

with boundary conditions

$$L_s^{(1)} \frac{\text{d}u_x}{\text{d}z}\bigg|_{z=0} = u_x(0) \ \text{ and } \ L_s^{(2)} \frac{\text{d}u_x}{\text{d}z}\bigg|_{z=h} = V_{\text{wall}} - u_x(h). \tag{5.69}$$

This boundary value problem solves to

$$u_x(z) = \frac{V_{\text{wall}}\left(z + L_s^{(1)}\right)}{h + L_s^{(1)} + L_s^{(2)}}. \tag{5.70}$$

The volumetric flow rate is readily found to

$$Q = \frac{V_{\text{wall}} w (h + 2L_s^{(1)}) h}{2(L_s^{(1)} + L_s^{(1)} + h)}, \tag{5.71}$$

and from this the flow enhancement is

$$E^{\text{slip}} = \frac{h + 2L_s^{(1)}}{h + L_s^{(1)} + L_s^{(2)}}. \tag{5.72}$$

We can see that the presence of slip can, in fact, reduce the flow rate for Couette flows because

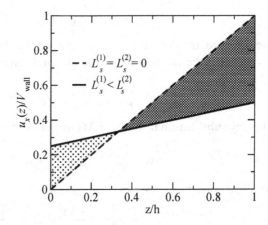

Figure 5.14 The Couette flow profile for different slip lengths. The shaded regions illustrate the volumetric flow rate difference for the two cases $L_s^{(1)} = L_s^{(2)} = 0$ (dashed line) and $L_s^{(1)} < L_s^{(2)}$ (full line).

$$L_s^{(1)} < L_s^{(2)} \text{ implies } E^{\text{slip}} < 1$$
$$L_s^{(1)} = L_s^{(2)} \text{ implies } E^{\text{slip}} = 1 \tag{5.73}$$
$$L_s^{(1)} > L_s^{(2)} \text{ implies } E^{\text{slip}} > 1.$$

A special case of the second condition is where there is no slip and we have $L_s^{(1)} = L_s^{(2)} = 0$, which is the standard macroscopic picture. The velocity profile for this situation is plotted in Fig. 5.14 together with the case where $L_s^{(1)} < L_s^{(2)}$. The two shaded regions illustrate the difference in flow rate, such that in the light shaded region the local flow rate is larger for case $L_s^{(1)} < L_s^{(2)}$, but (relatively) much smaller in the dark shaded region, leading to a net enhancement coefficient below unity.

5.4.3 Oscillatory Flows

The Poiseuille and Couette flows are steady flows. A non-steady flow can be realized by adding a time-varying pressure gradient or external force field of the form $\rho g \cos(\omega_0 t)$, where ρg is the force field amplitude and ω_0 the frequency – again, in our geometry the force acts in the x-direction. We will let the geometry be the same as for the planar Poisuille flow, and the Navier–Stokes equation becomes

$$\rho \frac{\partial u_x}{\partial t} = \rho g \cos(\omega_0 t) + \eta_0 \frac{\partial^2 u_x}{\partial z^2}. \tag{5.74}$$

The focus here will not be on the effect of the slip, and we let $L_s^{(1)} = L_s^{(2)} = 0$ for simplicity.

To proceed, we assume that the streaming velocity $u_x(z,t)$ can be separated into a product of a function of time and a function of space. For convenience, this is written in complex form,

$$u_x(z,t) = \text{Re}\left[U(z)e^{i\omega_0 t}\right], \tag{5.75}$$

and we can solve the differential equation in terms of the complex function $U(z)e^{i\omega_0 t}$, and u_x will then be the real part. The external force density can also be written in complex form,

$$\rho g \cos(\omega_0 t) = \text{Re}[\rho g e^{i\omega_0 t}], \tag{5.76}$$

and then by substitution into Eq. (5.74) we get the complex differential equation

$$i\rho \omega e^{i\omega_0 t} U(z) = \rho g e^{i\omega_0 t} + \eta_0 e^{i\omega_0 t} \frac{d^2 U(z)}{dz^2}, \tag{5.77}$$

which is rearranged to

$$\frac{d^2 U(z)}{dz^2} - \frac{i\omega_0 \rho}{\eta_0} U(z) = -\frac{\rho g}{\eta_0}. \tag{5.78}$$

Since the slip length is zero at both walls, the problem is solved using the boundary values $U(0) = U(h) = 0$.

Recall, the solution for this inhomogenous second-order differential equation is the sum of the particular solution U_p and the homogenous solution, U_h; that is,

$$U(z) = U_h(z) + U_p(z). \tag{5.79}$$

The eigenvalues for the homogenous solution is $\pm\sqrt{i\omega_0\rho/\eta_0}$, and using the identity $\sqrt{i} = \pm(1+i)/\sqrt{2}$, the homogenous solution is simply $U_h(z) = K_1 e^{(\alpha+i\alpha)z} + K_2 e^{-(\alpha+i\alpha)z}$, where

$$\alpha = \sqrt{\frac{\omega_0}{2v_0}}, \tag{5.80}$$

with $v_0 = \eta_0/\rho$ being the kinematic viscosity. K_1 and K_2 are integration constants determined by the boundary conditions.

The particular solution is easily found in this problem, as we simply guess that $U_p(z) = Az^2 + Bz + C$, giving $U_p = g/i\omega_0$ by inserting this solution into the differential equation, and compare the terms. This is known as the method of undetermined coefficients [34].[2]

Applying the no-slip boundary conditions, we obtain the solution for U,

$$U(z) = \frac{g}{i\omega}\left[1 - e^{(\alpha+i\alpha)z} - \left(1 - e^{(\alpha+i\alpha)h}\right)\frac{\sinh\left((\alpha+i\alpha)z\right)}{\sinh\left((\alpha+i\alpha)h\right)}\right]. \tag{5.81}$$

Before reaching the final expression for the streaming velocity, it is worthwhile to discuss the flow in terms of the Womersley number, Wo. This is defined as

$$\text{Wo} = \sqrt{\omega_0/v_0}h, \tag{5.82}$$

and, hence, can be written in terms of α. For liquid water flowing in a nanochannel of height 100 nm, the Womersley number is Wo $\approx \sqrt{\omega_0} \times 10^{-4}\text{s}^{1/2}$. Thus, for $\omega_0 = 10$ MHz, we have Wo ≈ 0.32, and for $\omega_0 = 1$ GHz, Wo ≈ 3.2, assuming that the viscosity is independent of the frequency. For nanoscale fluid systems at realistic frequencies,

---

[2] Appendix A.4 shows a few worked examples of how to use the method of undetermined coefficients.

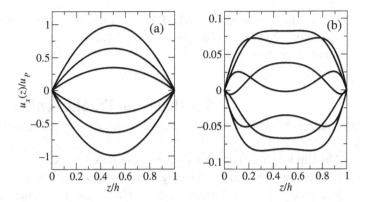

Figure 5.15 Flow profiles for water confined in a 100 nm slit-pore at different times. The streaming velocity is normalized with respect to the velocity at $z = h/2$ for the corresponding steady-state Poisuille flow, that is, where $\omega_0 = 0$. (a) $\omega_0 = 10\,\text{MHz}$. (b) $\omega_0 = 10\,\text{GHz}$.

the Womersley parameter is thus low, Wo < 1, and this motivates an expansion of the exponential and hyperbolic functions

$$e^{(\alpha+i\alpha)z} = 1 + (\alpha+i\alpha)z + \frac{1}{2}(\alpha+i\alpha)^2 z^2 + \dots \tag{5.83}$$

$$\sinh((\alpha+i\alpha)z) = (\alpha+i\alpha)z + \dots. \tag{5.84}$$

Substituting up to second order for the exponential function and to first order for the hyperbolic function, we arrive at the approximate solution for U,

$$U(z) \approx \frac{g\alpha^2}{\omega_0} z(h - z), \tag{5.85}$$

for sufficiently low Womersley number. The streaming velocity is then

$$u_x(z,t) = \text{Re}\left[U(z)e^{i\omega_0 t}\right] \approx \frac{\rho g}{2\eta_0} z(h - z)\cos(\omega_0 t). \tag{5.86}$$

This is simply a Poiseuille flow with a temporal oscillating factor.

Instantaneous profiles for two different Womserley numbers are shown in Fig. 5.15. The oscillatory Poiseuille-type flow is clearly seen for low Womersley number, whereas for large values of Wo the profiles behave in a plug-like manner. The plug flow is an indication of a decreasing boundary layer, leading to the definition of the Stokes boundary layer, δ_S, for oscillatory flows:

$$\delta_S = \frac{1}{\alpha} = \sqrt{\frac{2\nu_0}{\omega_0}}. \tag{5.87}$$

The Womersley number can then be written as Wo $= \sqrt{2}h/\delta_S$, which is less than unity for $\delta_S > \sqrt{2}h$; thus, for a slit-pore geometry, the Stokes layers may overlap in nanoscale geometries.

On a final note here, d-NEMD simulations have been carried out for nanoscale oscillatory flows [96, 178], showing excellent agreement between the theory and the simulations data.

5.5 Theory for the Slip Length

The slip length L_s can be estimated from flow rate experiments; however, these experiments are technical challenging, and the results are not always straightforward to interpret [125, 153]. Therefore, molecular dynamics simulations play (once more) a critical role when studying the slip phenomenon and thus when determining the slip length. One straightforward approach here is to perform d-NEMD simulations and calculate the streaming velocity profile. By application of Eq. (5.42) one can then find the slip length through fitting in the limit of small strain rates. This is a simple method which can always be attempted; however, for large slip lengths characterising plug-like flows, the values are associated with large statistical uncertainties [125].

To overcome this difficulty, different techniques for extracting the slip length using equilibrium simulations have been developed. The first attempt at such a theory is due to Bocquet and Barrat [23], who expressed the Navier friction coefficient ζ_N in terms of a Green–Kubo integral of the total shearing force in the fluid. This method is, however, debated [115, 171, 172], including by the authors themselves [24], showing a system size dependency and a non-convergence of the integral. Other approaches have followed, and these are typically based on modelling the fluctuating dynamics of either (i) the wall [24], (ii) a fluid volume element adjacent to the wall [105, 115], or (iii) the entire fluid volume (based on the original idea from Bocquet and Barrat) [191].

As an example, we discuss the second approach. The strategy here is to

1. Derive the streaming velocity profile for a particular flow – as we have already done – but using the average velocity of the interfacial fluid element $\mathcal{V}$ as the boundary condition.
2. Then, using a simple model for interfacial fluid velocity based on the wall–fluid and fluid–fluid frictional forces, we can express the entire system velocity profile in terms of a friction coefficient, and
3. from this get the expression for the slip length as a function of the friction coefficient.
4. Devise the method for calculating the friction coefficients.

We will chose a Couette flow with the usual geometry; see Fig. 5.16. The flow is realised by moving wall 2 with speed V_{wall}, keeping wall 1 stationary. Recall that for a Couette flow the Navier–Stokes equation reduces to a simple Laplace equation,

$$\frac{d^2 u_x}{dz^2} = 0. \qquad (5.88)$$

The boundary conditions are specified through the velocity of the interfacial fluid volume, here a slab of width Δ. Specifically, we use the spatial average slab velocity, giving rise to integral boundary conditions. At wall 1 we have

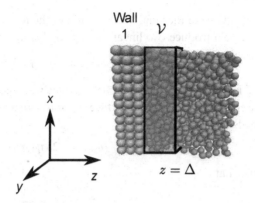

Figure 5.16
The fluid slab element with width Δ adjacent to the wall.

$$\overline{u}^{(1)} = \langle u_{\text{slab}} \rangle = \frac{1}{\Delta} \int_0^\Delta u_x(z) \, dz, \tag{5.89}$$

and at wall 2,

$$\overline{u}^{(2)} = V_{\text{wall}} - \langle u_{\text{slab}} \rangle = \frac{1}{\Delta} \int_{h-\Delta}^h u_x(z) \, dz. \tag{5.90}$$

Strictly, the average streaming velocity must be found from the average linear momentum in the slab. For example, for wall 1 it is assumed that

$$\frac{1}{m} \int_\mathcal{V} \rho u_x(z) \, d\mathbf{r} = \frac{1}{\Delta} \int_0^\Delta u_x(z) \, dz, \tag{5.91}$$

which is valid only if the density is constant, which we know is not the case, and we must be aware of this assumption.

Proceeding, the general solution to the Laplace equation is $u_x(z) = K_1 z + K_2$, where the constant of integration is found by the integrals, Eqs. (5.89) and (5.90). The solution to the problem is

$$u_x(z) = \frac{2\langle u_{\text{slab}} \rangle - V_{\text{wall}}}{\Delta - h} \left(z - \frac{\Delta}{2} \right) + \langle u_{\text{slab}} \rangle. \tag{5.92}$$

Notice that the strain rate $\dot{\gamma} = 1/2(du/dz)$ is, in terms of the slab velocity,

$$\dot{\gamma} = \frac{2\langle u_{\text{slab}} \rangle - V_{\text{wall}}}{2(\Delta - h)}. \tag{5.93}$$

Next, we seek a model for the average fluid slab-element velocity. For sufficiently large slab volume, the fluctuations of the mass in the slab can be ignored. The total force acting on the slab in the x-direction is given by two surface forces, namely the surface frictional force due to wall–fluid interactions, $F_{wf}(t)$, and a surface frictional force due to fluid–fluid interactions, $F_{ff}(t)$. The x-component of the centre-of-mass velocity, u_{slab}, is then written in terms of Newton's second law,

$$m \frac{du_{\text{slab}}}{dt} = F_{wf}(t) + F_{ff}(t). \tag{5.94}$$

In the steady state the time average follows the force balance, $\langle F_{wf} \rangle_t + \langle F_{ff} \rangle_t = 0$. To proceed, we introduce two linear models for F_{wf} and F_{ff}, namely,

$$\langle F_{wf} \rangle_t = -\zeta \langle u_{\text{slab}} \rangle_t \quad \text{and} \quad \langle F_{ff} \rangle_t = 2A\eta_0 \langle \dot{\gamma} \rangle_t, \tag{5.95}$$

where ζ is the friction coefficient between the wall and the slab. Note: in the first equation we effectively replace the surface frictional force with a volume force. Substitution then leads to

$$-\zeta \langle u_{\text{slab}} \rangle_t + 2A\eta_0 \langle \dot{\gamma} \rangle_t = 0, \tag{5.96}$$

implying that

$$\langle \dot{\gamma} \rangle_t = \frac{\zeta \langle u_{\text{slab}} \rangle_t}{2A\eta_0}. \tag{5.97}$$

Substitution of Eq. (5.97) into Eq. (5.93) gives an expression for the slab velocity in terms of the friction coefficient:

$$\langle u_{\text{slab}} \rangle_t = \frac{A\eta_0 u_{\text{wall}}}{2A\eta_0 - (\Delta - h)\zeta}. \tag{5.98}$$

Substituting into Eq. (5.92) and applying the Navier boundary condition,

$$L_s \frac{\partial u_x}{\partial z} \bigg|_{z=0} = u_x(0), \tag{5.99}$$

we obtain the final result for slip length,

$$L_s = \frac{A\eta_0}{\zeta} - \frac{\Delta}{2} = \frac{\eta_0}{\zeta_N} - \frac{\Delta}{2}, \tag{5.100}$$

where we have identified the relation $\zeta_N = \zeta/A$. As $\Delta \to 0$, we retrieve the Navier definition of the slip length.

The question still remains how to evaluate the friction coefficients ζ and η_0. The viscosity can simply be found from usual methods, for example, through the Green–Kubo integral, Eq. (3.38). The wall–fluid interfacial friction can be found from the generalized Langevin equation. If F_{wf} is the instantaneous force exerted by the wall on the fluid surface, we can write this as

$$F_{wf}(t) = -\int_0^t \zeta(t - t') u_{\text{slab}}(t')\, dt' + \delta F_{wf}(t), \tag{5.101}$$

where ζ is the generalized friction kernel analogous to the viscosity kernel in Chapter 4, and δF_{wf} is a random force with zero mean uncorrelated with the slab velocity, that is,

$$\langle \delta F_{wf}(t) u_{\text{slab}}(0) \rangle = 0. \tag{5.102}$$

Multiplying by $u_{\text{slab}}(0)$, averaging over an ensemble of initial conditions, and applying the linear properties if the integral, we arrive at an expression for the force–velocity correlation function,

$$\langle F_{wf}(t) u_{\text{slab}}(0) \rangle = C_{Fu}(t) = -\int_0^t \zeta(t - t') C_{uu}(t')\, dt', \tag{5.103}$$

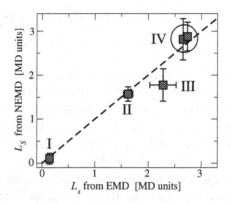

Figure 5.17 Comparison between the predicted slip length (x-axis) and results from direct-NEMD simulations (y-axis) for four different systems. The systems differ by their interaction potential between wall and fluid as well as densities and temperatures. The circle highlights two NEMD results from both Couette and Poiseuille flows. The fluids are Lennard–Jones type fluid systems; see details in Ref. [105].

in terms of the slab velocity autocorrelation function,

$$C_{uu}(t) = \langle u_{\text{slab}}(t)u_{\text{slab}}(0) \rangle. \tag{5.104}$$

A more convenient form is obtained by a Laplace transformation with respect to the generalized Laplace complex coordinate s. Applying the convolution theorem once again, we obtain

$$\widehat{C}_{Fu}(s) = \widehat{\zeta}(s)\widehat{C}_{uu}(s). \tag{5.105}$$

Both C_{Fu} and C_{uu} are easily evaluated from EMD simulations [105, 123], and through the Laplace transform the Navier friction coefficient can be found by taking the $s \to 0$ limit, that is, $\zeta = -\lim_{s \to 0} \widehat{C}_{Fu}(s)/\widehat{C}_{uu}(s)$. This leads to poor statistics, and the authors advised an alternative approach; see details in Ref. [209].

It is important to note that the slab height Δ should be approximately 1–2 particle diameters; for smaller heights, not all relevant wall–fluid interactions are included, and the friction coefficient will be underestimated. This differs from the original idea of Navier, where one considers the friction on the wall–fluid interface surface and not friction in a volume.

Fig. 5.17 shows the slip length obtained from d-NEMD simulations versus the predicted slip length, Eq. (5.100). Results from both Couette and planar Poiseuille flows show good agreement for all four different systems investigated, here indicated by roman numerals. Note that for these systems the slip length is relatively small and the equilibrium results can, without problems, be compared to d-NEMD simulations data. The method has also been successfully applied to water in carbon nanotubes [124, 125], Fig. 5.13.

5.6 Molecular Flows

We will start our discussion of molecular flows by solving the dynamical equations for the Couette and planar Poiseuille flows. The system geometries are shown in Figs. 5.12(b) and 5.8(a). However, in this exploration we will shift the coordinate system such that $z \in [-\overline{h}; \overline{h}]$, where $\overline{h}$ is half the slit-pore height, $\overline{h} = h/2$. While this may compromise the intuition of the problem a bit, the coordinate shift reduces the length of the resulting mathematical expressions considerably.

Since we now consider molecular fluids, we will allow the pressure tensor to have an antisymmetric part and we therefore include both the streaming velocity field $\mathbf{u}$ and spin angular velocity field $\mathbf{\Omega}$ as hydrodynamical variables. As usual, we focus on the low Reynolds number situation, as this is the case relevant for nanoscale flows. Then we ignore advection, viscous heating effects, and, finally, we assume zero divergence, that is, the usual incompressibility criterion $\nabla \cdot \mathbf{u} = 0$, but also $\nabla \cdot \mathbf{\Omega} = 0$.

The relevant balance equations are Eqs. (2.62) and (2.113), and the constitutive relations are given in Eqs. (3.22), (4.54), and (4.55). The corresponding divergence-free dynamical equations read

$$\rho \frac{\partial \mathbf{u}}{\partial t} = \rho \mathbf{g} + \eta_t \nabla^2 \mathbf{u} + 2\eta_r \nabla \times \mathbf{\Omega} \tag{5.106a}$$

$$\rho \Theta \frac{\partial \mathbf{\Omega}}{\partial t} = \rho \mathbf{\Gamma} + 2\eta_r (\nabla \times \mathbf{u} - 2\mathbf{\Omega}) + \zeta \nabla^2 \mathbf{\Omega}. \tag{5.106b}$$

Recall from Chapter 4 that $\eta_t = \eta_0 + \eta_r$ and $\zeta = \zeta_0 + \zeta_r$, and $\rho \mathbf{g}$ and $\rho \mathbf{\Gamma}$ are the applied force and torque densities.

5.6.1 The Molecular Couette Flow

The Couette flow is boundary driven and there are no external forces acting; that is, $\rho \mathbf{u}$ and $\rho \mathbf{\Gamma}$ are both zero. In the current geometry, only the y-component of the spin angular velocity field is non-zero. Then, for the steady-state flow regime, $u_x = u_x(z)$ and $\Omega_y = \Omega_y(z)$, we have

$$\eta_t \frac{d^2 u_x}{dz^2} - 2\eta_r \frac{d\Omega_y}{dz} = 0 \tag{5.107a}$$

$$2\eta_r \left(\frac{du_x}{dz} - 2\Omega_y \right) + \zeta \frac{d^2 \Omega_y}{dz^2} = 0. \tag{5.107b}$$

Here we will apply the Dirichlet no-slip boundary condition for both dynamical variables, that is, in our current coordinate system,

$$u_x(-\overline{h}) = 0, u_x(\overline{h}) = V_{\text{wall}} \text{ and } \Omega_y(-\overline{h}) = \Omega_y(\overline{h}) = 0. \tag{5.108}$$

Application of the no-slip boundary condition is, of course, a special case. A more general treatment will apply the Neumann boundary condition and include the possibility that these boundary conditions, too, are coupled. A discussion about the angular velocity boundary condition is made by Badur et al. [8]. For the points made here, the ad hoc no-slip condition suffices.

Proceeding, integration of Eq. (5.107a) gives

$$\frac{\mathrm{d}u_x}{\mathrm{d}z} = \frac{2\bar{\eta}_r}{\eta_t}\Omega_y + K_1, \tag{5.109}$$

where K_1 is a constant of integration. Substitution into Eq. (5.107b) yields an inhomogenous second-order differential equation for the angular velocity, namely,

$$\frac{\mathrm{d}^2\Omega_y}{\mathrm{d}z^2} + \frac{4\eta_r}{\zeta}\left(\frac{\eta_r}{\eta_t} - 1\right)\Omega_y + \frac{2\eta_r}{\zeta}K_1 = 0. \tag{5.110}$$

The general solution to this boundary value problem is found using the method of undetermined coefficients. The specific solution for the corresponding boundary value problem requires some algebra, and we here simply list the results; for the spin angular velocity,

$$\Omega_y(z) = \frac{2\eta_r\beta}{\alpha^2\zeta}\left[\frac{\cosh(\alpha z)}{\cosh(\alpha\bar{h})} - 1\right], \tag{5.111}$$

where α is the eigenvalue of the problem,

$$\alpha = 2\sqrt{\frac{\eta_r\eta_0}{\zeta\eta_t}}, \tag{5.112}$$

and where we have introduced the following auxiliary constants:

$$\beta = \frac{V_{\text{wall}}\alpha^3\eta_t\zeta}{8\eta_r^2\gamma - 2\bar{h}\alpha^3\eta_t\zeta} \quad \text{and} \quad \gamma = \tanh(\alpha\bar{h}) - \alpha\bar{h}. \tag{5.113}$$

The streaming velocity is found by integration of Eq. (5.109) to

$$u_x(z) = \beta\left[\frac{4\eta_r^2}{\eta_t\alpha^2\zeta}\left(\frac{\sinh(\alpha z)}{\alpha\cosh(\alpha\bar{h})} - z\right) - z\right] + \frac{V_{\text{wall}}}{2}. \tag{5.114}$$

The classical Couette flow solution in terms of the shifted coordinate, $\bar{h}$, is $u_x = V_{\text{wall}}(z/\bar{h} + 1)/2$. Also, for spin angular velocity we have $\Omega_y = (1/2)\mathrm{d}u_x/\mathrm{d}z = V_{\text{wall}}/(4\bar{h})$. In particular, as the spin angular velocity is given directly from the streaming velocity, the classical theory does not allow for boundary specifications, and the extended and classical predictions will differ by a magnitude of $V_{\text{wall}}/(4\bar{h})$ at $z \pm \bar{h}$. For sufficiently high channels and as we approach the channel mid-point, we will expect that the extended theory converges to that of the classical theory.

Figure 5.18 shows the velocity and spin angular velocity profiles obtained from a d-NEMD simulation of a butane fluid undergoing a Couette flow. The streaming velocity profile is calculated from the molecular centre-of-mass linear momentum, Eq. (5.47), and the local angular velocity is calculated using its definition, Eq. (2.99), noting that the time-averaged spin angular momentum and momentum of inertia are

$$\mathbf{S}(z) = \frac{\langle\sum_i \mathbf{S}_i\delta(z - z_i)\rangle_t}{\langle\sum_i m_i\delta(z - z_i)\rangle_t} \quad \text{and} \quad \mathbf{\Theta}(z) = \frac{\langle\sum_i m_i\mathbf{\Theta}_i\delta(z - z_i)\rangle_t}{\langle\sum_i m_i\delta(z - z_i)\rangle_t}. \tag{5.115}$$

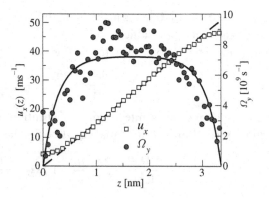

Figure 5.18 The streaming velocity and spin angular velocity profiles for a model butane fluid undergoing a Couette flow. Symbols are results from d-NEMD simulations, and lines are predictions (no fitting) from hydrodynamic theory using $\eta_0 = 0.14$ mPa·s, $\eta_r = 0.013$ mPa·s, and $\zeta = 4 \times 10^{-24}$ kg ms^{-1}, Table 4.1. The bulk density is approximately 582 kgm^{-3} and temperature is $T = 288$ K.

The simulation results are accompanied with the hydrodynamic predictions. The agreement is satisfactory away from the wall, but at the wall–fluid interface the predicted velocity profile does not agree with the molecular dynamics results. As we saw in Figs. 5.2 and 5.4, the system features strong density variation and molecular alignment in the wall–fluid interface. Our hydrodynamical model here assumes both homogeneity and isotropy, which is, again, too crude an assumption near the wall, and we expect to observe deviations.

Importantly, the spin angular velocity is much lower than predicted by the classical hydrodynamic theory, $V_{\text{wall}}/(4\bar{h})$, which is on the order of 10^{11} s^{-1}.

It is highly informative to study the corresponding symmetric and antisymmetric shear pressures. They are given directly from the constitutive relations,

$$\overset{os}{P}_{xz}(z) = -\eta_0 \frac{\mathrm{d}u_x}{\mathrm{d}z} \quad \text{and} \quad \overset{ad}{P}_y(z) = -\eta_r \left(\frac{\mathrm{d}u_x}{\mathrm{d}z} - 2\Omega_y \right), \tag{5.116}$$

and plotted in Fig. 5.19. The classical hydrodynamic theory predicts a constant symmetric shear pressure, $\overset{os}{P}_{xz} = \eta_0 V_{\text{wall}}/(2\bar{h})$. From Eqs. (5.109) and (5.114) it is easily seen that the extended theory predicts that $\overset{os}{P}_{xz}$ varies with the spatial coordinate. This is the result of the coupling and is significant at the wall–fluid interface. This point can also be concluded from the antisymmetric shear pressure, as this features the same qualitative behaviour. The classical hydrodynamic theory predicts a zero antisymmetric shear pressure, which is only observed sufficiently far away from the wall–fluid interface.

5.6.2 The Molecular Poiseuille Flow

We continue and explore the planar Poiseuille flow. The extended hydrodynamic equations are

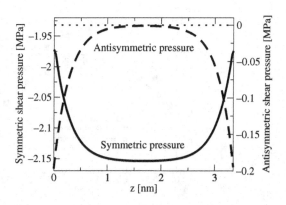

Figure 5.19 The symmetric (full line) and antisymmetric (broken line) shear pressure profiles for butane fluid undergoing a Couette flow. Notice the two difference scales on the y-axis. The horizontal dots indicate zero antisymmetric stress. The transport coefficients are given in Fig. 5.18.

$$\eta_t \frac{d^2 u_x}{dz^2} - 2\eta_r \frac{d\Omega_y}{dz} = -\rho F_e \qquad (5.117a)$$

$$2\eta_r \left(\frac{du_x}{dz} - 2\Omega_y \right) + \zeta \frac{d^2\Omega_y}{dz^2} = 0, \qquad (5.117b)$$

where we let the flow be generated by an external gravitational-like force. To compare with data later, we let the boundary conditions be simple Dirichlet boundaries,

$$u_x(-\overline{h}) = u_x(\overline{h}) = u_w \text{ and } \Omega_y(-\overline{h}) = \Omega_y(\overline{h}) = 0. \qquad (5.118)$$

The slip-velocity u_w needs to be determined experimentally or from simulations.

The solution to this boundary value problem can be found using the same approach as previously. Integrating Eq. (5.117a),

$$\frac{du_x}{dz} = \frac{2\eta_r}{\eta_t} \Omega_y - \frac{\rho F_e}{\eta_t} z + K_1, \qquad (5.119)$$

which gives the differential equation for the spin angular velocity,

$$\frac{d^2\Omega_y}{dz^2} + \frac{4\eta_r}{\zeta} \left(\frac{\eta_r}{\eta_t} - 1 \right) \Omega_y - \frac{2\eta_r \rho F_e}{\eta_t \zeta} z + \frac{2\eta_r}{\zeta} K_1 = 0. \qquad (5.120)$$

Again, using the method of undetermined coefficients, we arrive, after some algebra, at

$$\Omega_y(z) = \frac{\rho F_e \overline{h}}{2\eta_0} \left[\frac{z}{\overline{h}} - \frac{\sinh(\alpha z)}{\sinh(\alpha \overline{h})} \right] \qquad (5.121)$$

for spin angular velocity, and

$$u_x(z) = \frac{\rho F_e \overline{h}^2}{2\eta_0} \left[1 - \left(\frac{z}{\overline{h}} \right)^2 + \frac{2\eta_r \coth(\alpha \overline{h})}{\eta_t \alpha \overline{h}} \left(\frac{\cosh(\alpha z)}{\cosh(\alpha \overline{h})} - 1 \right) \right] + u_w \qquad (5.122)$$

for the streaming velocity. See also Eringen, Ref. [62].

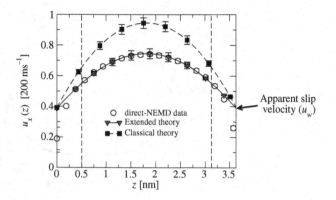

Figure 5.20   The streaming velocity profile for water flowing in a slit-pore. Open circles are results from d-NEMD, filled squares connected with lines represent the classical hydrodynamic predictions, and the triangle the extended hydrodynamic predictions. From Ref. [105] with permission.

The corresponding classical hydrodynamic result is found in the case of $\eta_r = 0$; that is, we get the velocity,

$$u_x(z) = \frac{\rho F_e \bar{h}^2}{2\eta_0} \left[1 - \left(\frac{z}{\bar{h}} \right)^2 \right] + u_w, \tag{5.123}$$

and the corresponding spin angular velocity,

$$\Omega_y(z) = -\frac{\rho F_e}{\eta_0} z. \tag{5.124}$$

Figure 5.20 plots the streaming velocity profile obtained from d-NEMD simulations of water undergoing a planar Poiseuille flow in a slit-pore with a height of 3.6 nm [105]. To compare the predictions with simulation results, we need to extract the slip velocity, u_w. If the velocity profile is extrapolated only using the velocity profile away from the (troublesome) wall–fluid interface, an *apparent* slip velocity (or equivalently an apparent slip length) can be inferred. This apparent slip can defined from both the classical and extended theories, Eqs. (5.123) and (5.122). Both predictions are plotted as symbols connected with lines. Clearly, the extended theory is in good agreement with the simulation data, whereas the classical theory predicts too large a flow.

From Fig. 5.20 it is evident that the volumetric flow rate is significantly reduced because of the coupling. The effect can be quantified through the enhancement coefficient $E^{\text{coup}} = Q^{\text{extend}}/Q^{\text{class}}$. By integration, we arrive at [102]

$$E^{\text{coup}} = 1 + \frac{3\eta_r \gamma}{\eta_t \tanh(\alpha \bar{h})(\alpha \bar{h})^2}. \tag{5.125}$$

γ is given in Eq. (5.113). The second term on the right-hand side goes to zero as $\alpha \bar{h}$ increases. Therefore, for large channel heights the flow rate predicted by the classical theory is the same as that predicted by the extended theory, which is also expected.

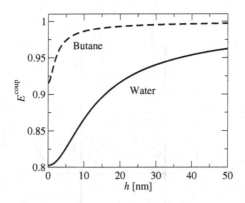

Figure 5.21 Coupling enhancement coefficient, Eq. (5.125), as a function of channel height for butane and water fluids. Parameter values for η_r, η_0, and ζ can be found in Table 4.1.

In order to quantify the effect of the coupling in terms of a length we introduce [93]

$$l_c = \sqrt{\zeta/\eta_r},\tag{5.126}$$

such that the fundamental eigenvalue for the problem can be written as

$$\alpha = \frac{2}{l_c}\sqrt{\frac{\eta_0}{\eta_t}} \approx \frac{2}{l_c},\tag{5.127}$$

since $\eta_0 > \eta_r$ for most fluids. Then, large l_c, indicates large effect of the coupling on the flow enhancement due to the additional dissipative process coming from spin angular velocity diffusion. From Table 4.1, we have for the butane fluid $l_c \approx 0.5$ nm and for liquid water $l_c \approx 3.5$ nm; hence, the coupling effect is small for butane compared to water. This is seen in Fig. 5.21, where the enhancement coefficient is plotted as a function of channel height for the two fluids.

5.6.3 Torque Insertion

We now consider the situation where the external torque density is non-zero. The hydrodynamic equations for the slit-pore are

$$\eta_t \frac{\mathrm{d}^2 u_x}{\mathrm{d}z^2} - 2\eta_r \frac{\mathrm{d}\Omega_y}{\mathrm{d}z} = 0\tag{5.128a}$$

$$2\eta_r \left(\frac{\mathrm{d}u_x}{\mathrm{d}z} - 2\Omega_y\right) + \zeta_t \frac{\mathrm{d}^2 \Omega_y}{\mathrm{d}z^2} = -\rho\Gamma.\tag{5.128b}$$

In this particular example, we let the boundary conditions be mixed in that we allow for a velocity slip at one wall:

$$u_x(-\overline{h}) = 0, \;\; L_s \left.\frac{\partial u_x}{\partial z}\right|_{z=\overline{h}} = -u_x(\overline{h}) \;\; \text{and} \;\; \Omega_y(-\overline{h}) = \Omega_y(\overline{h}) = 0.\tag{5.129}$$

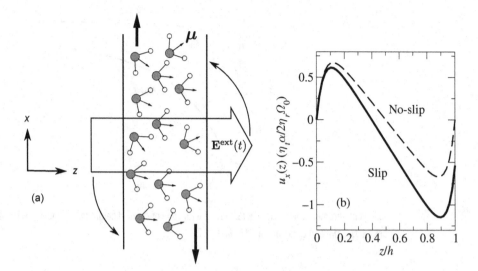

Figure 5.22 (a) Schematic illustration of an external rotational electric field exerting a torque on water confined in a slit-pore. From Ref. [92] with permission. (b) Examples of the resulting velocity profiles for no-slip and slip situations. The velocity and spatial coordinates are normalized.

The differential equation for the spin angular velocity is readily found to be

$$\frac{d^2\Omega_y}{dz^2} + \frac{4\eta_r}{\zeta}\left(\frac{\eta_r}{\eta_t} - 1\right)\Omega_y + \frac{2\eta_r + \rho\Gamma}{\zeta}K_1 = 0. \tag{5.130}$$

and is seen to be on the same functional form as the equation for the Couette flow. We therefore write the general solution as

$$\Omega_y(z) = \Omega_0\left[\frac{\cosh(\alpha z)}{\cosh(\alpha\bar{h})} - 1\right], \tag{5.131}$$

where Ω_0 is the spin velocity amplitude depending on the applied torque. Integrating Eq. (5.128a) and applying the boundary conditions, we obtain for the streaming velocity

$$u_x(z) = \frac{2\eta_r\Omega_0}{\eta_t\alpha}\left[\frac{\sinh(\alpha\bar{z})}{\cosh(\alpha\bar{h})} - \left(\alpha + \frac{2\gamma}{L_s + 2\bar{h}}\right)z + \left(1 - \frac{2\bar{h}}{L_s + 2\bar{h}}\right)\gamma\right], \tag{5.132}$$

where, recall, $\gamma = \tanh(\alpha\bar{h}) - \alpha\bar{h}$.

For dielectric materials, the torque insertion can be realised by application of an external electric field [28], $\mathbf{E}^{\text{ext}}$; see Fig. 5.22. This is, in particular, relevant for nanoscale geometries, as large fields can be achieved with realistic experimental electric potential differences. The external electric field induces a polarisation $\mathbf{P}$, which will align with the field direction. The polarisation, in turn, leads to a screening field, and the local field $\mathbf{E}$ is a sum of the external field and this screening field; the local electric field gives raise to the torque density,

$$\rho\Gamma = \mathbf{P} \times \mathbf{E}. \tag{5.133}$$

Bonthuis et al. [28] showed through perturbation analysis that the torque is to first order in the electric field given by the external field, making the theoretical treatment tractable. Then, let the external electric field be given by

$$\mathbf{E}^{\text{ext}} = E_0 \sin(\omega_0 t)\mathbf{i} + E_0 \cos(\omega_0 t)\mathbf{k}, \tag{5.134}$$

where $\mathbf{i}$ and $\mathbf{k}$ are the unit vectors parallel to the x and z axes, respectively. If we further let the polarisation be homogenous and assume that we can ignore the non-linear cross coupling between the spin angular velocity and polarisation and we have the Debye equation, Eq. (2.125),

$$\frac{d\mathbf{P}}{dt} = \frac{1}{\tau_D} \left(\varepsilon_0 \chi_e \mathbf{E}^{\text{ext}} - \mathbf{P} \right). \tag{5.135}$$

In the large time limit where transient features are completed, the Debye equation solves to

$$\mathbf{P}(t) = \frac{\rho \varepsilon_0 \chi_e E_0}{1 + \omega_0^2 \tau_D^2} \left[(\cos(\omega_0 t) + \omega_0 \tau_D \sin(\omega_0 t))\mathbf{i} + (\sin(\omega_0 t) - \omega_0 \tau_D \cos(\omega_0 t))\mathbf{k} \right]. \tag{5.136}$$

The torque density only has one non-zero component, namely the y-component, which can be found directly from Eq. (5.133), where the local field $\mathbf{E}$ is replaced by the external field, giving

$$\rho\Gamma = -\frac{\rho \varepsilon_0 \chi_e E_0^2 \omega_0 \tau_D}{1 + \omega_0^2 \tau_D^2}. \tag{5.137}$$

From this we see that the torque density depends on the external field frequency and has a maximum at $\omega_0 = 2\pi/\tau_D$; for water, this corresponds to an angular frequency of approximately 0.125 THz.

Figure 5.22 plots two velocity profiles for a generic dielectric fluid exposed to a rotating field giving rise to a torque: one profile for $L_s > 0$, and one for $L_s = 0$. It is clear that for $L_s = 0$, the volumetric flow rate is zero, whereas it is non-zero for $L_s > 0$. Thus, if the slip conditions are asymmetric, the torque insertion method can be used as a pumping mechanism.

5.7 Further Explorations

1. Simulate a Couette flow for a confined Lennard–Jones liquid, where the channel height is around 10 Lennard–Jones particle diameters. The fluid state point can be $(\rho, T) = (0.75, 1.0)$. Change the wall–fluid interaction parameter (see computational resource); how does this affect the fluid flow velocity and density profiles.

   From the velocity profiles, calculate the slip lengths and compare the simulation data with the hydrodynamic prediction, Eq. (5.70).

Calculate the hydrodynamic channel height, Eq. (5.55), and compare this length with the non-zero density profile height.

Computational resources available.

2. An oscillatory Couette flow can be achieved by a periodic shear motion of one or both walls in a slit-pore. Assume the no-slip boundary conditions $u_x(0) = 0$ and $u_x(h) = V_{wall}$, where

$$V_{wall}(t) = V_0 \sin(\omega_0 t),$$

ω_0 being the imposed shear oscillation frequency. Solve the Navier–Stokes equation for this case.

To compare with simulation data, we take the time average over a small interval of the oscillation period $T_0 = 2\pi/\omega_0$. For example, we can split the period into eight time intervals such that the time average for the first interval is

$$\langle u_x \rangle_t = \frac{8}{T_0} \int_0^{T_0/8} u_x(z,t) \, dt \,.$$

Integrating from $T_0/8$ to $T_0/4$, we obtain the average for the second interval, and so forth. Derive the time averages for all eight intervals. Compare with molecular dynamics simulation data for different frequencies ω_0; tweak the interaction parameter between the wall and fluid particles to reduce the slip length as much as possible. How do you characterise the flow profile for different frequencies?

Computational resources available.

3. Derive the symmetric and antisymmetric shear pressure for a molecular Poiseuille flow. Use the values for the transport coefficients given in Table 4.1 to compare the pressure profiles for chlorine, butane, and water.

4. Prove that for no-slip boundary conditions, $u_x(0) = u_x(h) = 0$ and $\Omega_y(0) = \Omega_y(h) = 0$, torque insertion cannot result in an average channel flow.

6 Gradients

The previous chapter dealt with simple nanoscale fluid flows that were characterised by non-zero gradients in the fluid velocity. In this chapter we explore the formation and effects of gradients in temperature, electric potential, and chemical concentrations, and we will see new phenomena arise and how they are modelled using the hydrodynamic framework.

6.1 Phenomena in Temperature Gradients

In nanoscale systems extremely large temperature gradients can be achieved. For example, Fedoruk et al. [70] heated a nano-gold particle with an optical laser, obtaining a temperature gradient in the fluid surrounding the particle on the order of 1 K/nm. See also Govorov et al. [82] for a theoretical treatment of this system. Such extreme gradients enable us to investigate phenomena which cannot, at least so easily, be observed for macroscopic gradients.

Gradients in temperature give rise to a whole family of coupling phenomena; we will explore some of the more well-known ones here. We will also see examples that are perhaps not so commonly discussed in the literature, yet they do provide new insight into the hydrodynamics of fluids. The purpose of this section is to give an introduction to the rich dynamical phenomenology, and the reader is referred to the references for more details.

6.1.1 The Kapitza Length

It is enlightening to first explore a simple system composed of single-type non-charged molecules with an imposed constant temperature gradient. We envision that such a system can be realised in a slit-pore and by cooling one wall and heating the other, such that the temperatures are T_C and T_H, respectively; see Fig. 6.1. As a first approximation, we use the standard macroscopic picture and we expect the gradient to be

$$\alpha = (T_H - T_C)/h. \tag{6.1}$$

This need not be true. In particular, the temperature gradient can be smaller.

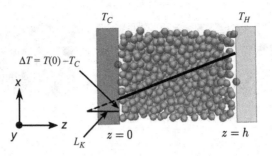

Figure 6.1 Illustration of the interfacial temperature jump and the Kapitza length. The linear temperature profile is shown as a black line.

To investigate this claim, we first assume constant density; we have from Eqs. (2.90), (2.87), and (3.22c) the temperature,

$$\frac{\partial T}{\partial t} = \lambda \frac{\partial^2 T}{\partial z^2},$$ (6.2)

for the slit-pore geometry. Note: as in Chapter 5, we here do not write the subscript "av" explicitly unless we study the actual fluctuations. This is the one-dimensional heat equation. The boundary conditions must be specified, and a first reasonable guess would be the Dirichlet boundaries $T(0) = T_C$ and $T(h) = T_H$, giving $T(z) = \alpha z + T_C$ and hence the expected gradient, Eq. (6.1).

Kapitza [126] showed that at the wall–fluid boundary there exists an abrupt change in thermal resistance (inversely proportional to the heat conduction). This resistance results in a temperature jump at the interface, for example, for the cold wall $\Delta T = T(0) - T_C$ as illustrated in Fig. 6.1.

Inspired by the Navier boundary condition discussed in Chapter 5, a linear extrapolation of the temperature profile from the wall–fluid boundary to the point where the extrapolation equals the wall temperature defines a length, L_K. This length is denoted the Kapitza length and we have

$$T(0) - T_C = L_K \left. \frac{\partial T}{\partial z} \right|_{z=0} \quad \text{and} \quad T(h) - T_H = -L_K \left. \frac{\partial T}{\partial z} \right|_{z=h}.$$ (6.3)

In general, this is written as the Neumann boundary,

$$\Delta T = L_K \nabla T \cdot \mathbf{n},$$ (6.4)

such that $\mathbf{n}$ is the wall normal vector pointing into the fluid; also see Fig. 5.8(b). In the steady state we integrate the right-hand side of Eq. (6.2), and application of the boundaries Eq. (6.3) yields

$$T(z) = \frac{T_H - T_C}{h + 2L_K} (z + L_K) + T_C.$$ (6.5)

Notice that the presence of an interfacial thermal resistance jump gives rise to a reduced system temperature gradient.

Table 6.1 Kapitza lengths for three different systems. σ is the usual Lennard–Jones length scale. The large interval for the FCC - LJ system is due to different wall–fluid interactions, see Ref. [7] for details. [a] Ref. [7] [b] Ref. [173]

Wall	Fluid	Kapitza length, L_K
[a] LJ - FCC	LJ (fluid)	5–30 σ
[b] Au - FCC	Water (liquid)	6.5 ± 0.7 nm
[b] Silicon - diamond	Water (liquid)	7.5 ± 0.3 nm

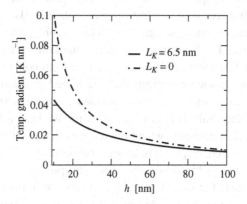

Figure 6.2 Temperature gradient as function of slit-pore height, where $T_H - T_C = 1$ K and $L_K = 6.5$ nm or $L_K = 0$.

In Table 6.1 L_K is listed for a few systems. Like the slip length, the Kapitza length depends on the detailed wall–fluid interactions, temperature, fluid density, and so on [69]. The general picture is that L_K decreases as the adhesive interactions increase, that is, as the attractive interactions between the wall and fluid atoms increase. It is worth noting that the Kapitza length can be evaluated through NEMD or EMD simulations [7, 14].

For a given Kapitza length and wall temperatures, the solution approaches

$$T(z) = \alpha z + T_C \tag{6.6}$$

as h increases. Figure 6.2 plots the temperature gradient as a function of slit-pore height h using $L_K = 6.5$ nm (the Au–Water system) and $L_K = 0$. Thus, for a pore height of 10 nm, the gradient is more than half of what we expect from the classical boundary condition.

Equation (6.2) is based on a local linear constitutive relation, specifically, Fourier's law. Away from the wall–fluid interface, this simple model suffices even for relatively large temperature gradients. However, in the interface region the different inter-facial phenomena (density variations, alignment, etc.) affect the transport properties, and a non-local anisotropic description is needed. Such a careful treatment is not yet available in the literature.

6.1.2 The Soret Effect

In the nineteenth century C. Ludwig and C. Soret independently discovered that an imposed temperature gradient in fluid mixtures induces a concentration gradient in the components. According to Platten, Ref. [175], it was Soret who made a more detailed study, and the phenomenon is therefore often referred to as the Soret effect. It is also known as thermodiffusion, thermomigration, and the Ludwig–Soret effect to commemorate Ludwig. Some authors use the term thermophoresis; however, this is also used specifically for colloidal suspensions, which we do not treat here.

One consequence of the Onsager reciprocal relation [52] is that if a temperature gradient induces a concentration gradient, it is also true that a concentration gradient induces a temperature gradient. This coupling must, of course, also agree with Curie's principle. The latter phenomenon is the Dufour effect. Thus, to model this phenomenon, we strictly need the equations for both concentrations and temperature, forming a set of coupled equations that are solved simultaneously. This proves to be a rather complicated exercise, as the problem is non-linear, and, for example, standard perturbation analysis is not very helpful. We here follow Miller et al. [159] and take a simplified approach, where the temperature profile is given a priori. From this we can derive the corresponding concentration profile. While this is, perhaps, less elegant, it still allows us to address the important points.

We base the modelling of the phenomenon on the single-particle dynamics. Recall Eq. (3.90); we here highlight that we explore the single-particle dynamics of a specific type of molecule, say type A, and ignoring advection, we get

$$\frac{\partial \rho_A}{\partial t} = -\boldsymbol{\nabla} \cdot \mathbf{J}^A. \tag{6.7}$$

Next, we need a constitutive relation for the particle flux $\mathbf{J}^A$. Following the standard formalism, this is expressed through the mass fraction, $x_A = \rho_A/\rho$, rather than ρ_A, and the flux is given by [52],

$$\mathbf{J}^A = -\rho D_{AB} \boldsymbol{\nabla} x_A - \rho D_T x_A (1 - x_A) \boldsymbol{\nabla} T, \tag{6.8}$$

again omitting the subscript av. D_{AB} is the mutual diffusion coefficient describing the diffusion of A in a system with a specific composition of A and B molecules; hence, the mutual diffusion coefficient is dependent on both the mixture's composition and specific state point. D_T is the thermal diffusion coefficient for A and is also mixture and state-point dependent. The symbol for the thermal diffusion coefficient should not be confused with the thermal diffusivity from Chapter 3.

Equation (6.8) includes the coupling between the temperature gradient and the mass flux in accordance with Curie's principle; notice that the coupling term is non-linear. Substitution of Eq. (6.8) into Eq. (6.7) yields the dynamical equation for the mass fraction,

$$\frac{\partial x_A}{\partial t} = \boldsymbol{\nabla} \cdot \left(D_{AB} \boldsymbol{\nabla} x_A + D_T x_A (1 - x_A) \boldsymbol{\nabla} T \right). \tag{6.9}$$

Table 6.2 Soret coefficient for different mixtures. For the water mixtures, the interval indicates the mixture's composition with respect to the organic compound. Parentheses indicate uncertainty on this digit. [a] Ref. [167] [b] Refs. [167, 218] [c] Ref. [159]

Mixture	Composition	S_T [10^{-3} K^{-1}]
Water–methanol[a]	$x_{\text{methanol}} = [0.0125; 0.64]$	3.(8) to −3.(2)
Water–ethanol[b]	$x_{\text{ethanol}} = [0.01; 0.2]$	8.(4) to −4.(7)
Water–acetone[a]	$x_{\text{acetone}} = [0.027; 0.903]$	6.(7) to −1.(1)
Krypton–argon[c]	Equimolar	13

In the following we focus on the steady state and let the fluid be confined between two parallel surfaces with different temperatures T_C and T_H, $T_C < T_H$, such that the direction of confinement is the z-direction; this is analogous to the situation in Fig. 6.1. Notice that, in accordance with the previous section, we do not demand that $T(0) = T_C$ and $T(h) = T_H$. In the steady state we have that the mass fraction and temperature are dependent only on the spatial z coordinate, $x_A = x_A(z)$ and $T = T(z)$. The advection is zero and we ignore thermal expansion, such that we explore only sufficiently small temperature differences. In the steady state we also have that the mass flux of A is zero everywhere, $\mathbf{J}^A = \mathbf{0}$. In the current geometry, we get

$$\frac{\mathrm{d}x_A}{\mathrm{d}z} = -S_T x_A (1 - x_A) \frac{\mathrm{d}T}{\mathrm{d}z}, \tag{6.10}$$

where $S_T = D_T / D_{AB}$ is the Soret coefficient. This equation is separable,

$$\frac{\mathrm{d}x_A}{x_A(1 - x_A)} = -S_T \mathrm{d}T, \tag{6.11}$$

and integration from $z' = 0$ to $z' = z$,

$$\int_{x_A(0)}^{x_A(z)} \frac{\mathrm{d}x_A}{x_A(1 - x_A)} = -S_T \int_{T(0)}^{T(z)} \mathrm{d}T, \tag{6.12}$$

yields, after manipulation,

$$\frac{1}{x_A(z)} = 1 + \frac{1 - x_A(0)}{x_A(0)} e^{S_T(T(z) - T(0))} \tag{6.13}$$

for temperature profile $T(z)$.

The Soret coefficient for four different mixtures is listed in Table 6.2. Notice that S_T can take both positive and negative values; a positive Soret coefficient means that the molecule type we investigate, here a generic type A, migrates down the temperature gradient; and vice versa, for negative values the molecules migrate up the temperature gradient.

As a first approach, we let the temperature be given by Eq. (6.5), which we write in the more compact form,

$$T(z) = \alpha_{L_K}(z + L_K) + T_C. \tag{6.14}$$

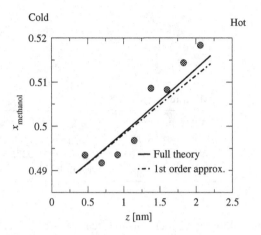

Figure 6.3 Concentration profile for methanol. The system is a water–methanol mixture, $x_{\text{methanol}} = 0.505$, and using synthetic NEMD the system is held at a constant temperature gradient of around 33 K/nm. Data are from Nieto-Draghi et al. [167]. Lines are graphs of Eqs. (6.13) and (6.16) using the Soret coefficient $2.5 \times 10^{-3}\,\text{K}^{-1}$.

where $\alpha_{L_K} = (T_H - T_C)/(h + 2L_K)$. Substituting of this into Eq. (6.13) and Taylor expanding around $z = 0$, we get

$$\frac{1}{x_A} = 1 + \frac{1 - x_A(0)}{x_A(0)}\, e^{S_T(\alpha_{L_K}(z - L_K) - \Delta T)}$$

$$= 1 + \frac{1 - x_A(0)}{x_A(0)}\, e^{S_T \alpha_{L_K} L_K}(1 + S_T \alpha_{L_K} z) + \cdots, \qquad (6.15)$$

for sufficiently small temperature jumps ΔT. For zero Kapitza length, the first-order Taylor expansion reduces to

$$\frac{1}{x_A} = 1 + \frac{1 - x_A(0)}{x_A(0)}\left(1 + \frac{S_T(T_H - T_C)}{h} z\right), \qquad (6.16)$$

hence, the prefactor $e^{S_T \alpha_{L_K} L_K}$ is a measure for the effect of the wall–fluid thermal resistance jump on the concentration. For example, using $L_K = 10$ nm and $S_T = 0.01\,\text{K}^{-1}$, we get a prefactor of 1.0033 for $h = 10$ nm and $T_H - T_C = 1$ K; that is, the effect from the interfacial thermal resistance on the concentration profile is very small.

Figure 6.3 plots simulation data for the concentration profile for methanol in a water–methanol mixture. The temperature gradient is constant at approximately 33 K/nm and held fixed through a synthetic technique. The theoretical predictions, Eqs. (6.13) and (6.16), are shown as lines.

6.1.3 Charged Systems in a Temperature Gradient

For electrolyte solutions, a thermal gradient induces an electrical potential gradient (or voltage). This phenomenon is known as the thermoelectric effect. Also, by imposing an electric potential gradient, one induces a temperature gradient. If φ_q is the electric potential and σ the electric conductivity, the linear constitutive relation for the charge flux, that is, the electric current, reads [131]

$$\mathbf{j}_e = -\sigma\nabla\varphi_q - \sigma S_e\nabla T, \tag{6.17}$$

where S_e is the Seebeck coefficient. Like the Soret coefficient, the Seebeck coefficient can take negative values; for alkali-halide electrolyte solutions, Lecce and Bresme [144] found that -1.5×10^{-4}V K^{-1} $< S_e < 3\times10^{-4}$V K^{-1} over a large range of temperatures and salt concentrations.

Consider the system in Fig. 6.1. If the walls are not electrodes capable of conducting charges, the confinement prevents a constant electrical current. Therefore, for sufficiently large times, $\mathbf{j}_e = 0$ and, if we further assume that the gradients are constants, we have from Eq. (6.17) the potential difference across the slit-pore,

$$\Delta\varphi_q = -S_e\Delta T. \tag{6.18}$$

Using $h = 4$ nm, and for later comparison a fictitious large temperature difference $\Delta T = 132$K (a gradient of 33 K nm^{-1}), and $S_e = 10^{-4}$V K^{-1}, the potential difference will be on the order of 10^{-2} V, corresponding to a very large electric field of 3.3×10^6 Vm^{-1}. This crude estimate not only ignores the temperature dependence of the Seebeck coefficient, it also does not include screening and layering effects, and other phenomena.

A more careful calculation of the local electric field can be done by application of Gauss' law,

$$E(z) = \frac{1}{\varepsilon_0}\int_0^z \rho_q(z')\mathrm{d}z', \tag{6.19}$$

where ρ_q is the total charge density. Figure 6.4 re-plots data from molecular dynamics simulations for the local electric field from Lecce and Bresme [144]. We see that the crude estimate for the electric field above yields the right order of magnitude, but it naturally fails as a direct quantitative prediction. In the next section, we return to charged systems and explore the layering and screening in a more systematic manner.

Finally, it is interesting to note that a temperature gradient also induces popularisation, leading to so-called thermopolarisation. Using molecular dynamics simulations of water applying a large temperature gradient on the order of 1 K/nm, Bresme et al. [35] were able to produce a significant non-zero polarisation, and equivalently a non-zero local electric field.

6.1.4 Fluctuations in a Temperature Gradient

In Chapters 3 and 4 we studied the dynamics through fluctuations in equilibrium. We push this idea further and here explore the fluctuations under non-equilibrium conditions; and a new surprising coupling emerges.

Inspired by Kirkpatrick et al. [129, 130] and Zárate and Sengers [56], we continue to apply the geometry given in Fig. 6.1. To keep it simple yet still to the point, the Kapitza length is zero, $L_K = 0$. Then, at $z = 0$, the temperature is T_C, and at $z = h$, it is

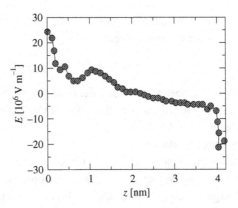

Figure 6.4 The resulting electric field profile for a LiCl–water solution (1.0 mol kg^{-1}) when applying a temperature gradient of 33 K nm^{-1}. Data are re-plotted from Lecce and Bresme [144]; only part of the system electric field profile is shown.

T_H. The fluid is composed of uncharged molecules all of the same type, so we ignore the Soret, thermoelectric, and thermopolarization effects. We will exclude external forces, and therefore $\mathbf{u}_{av} = \mathbf{0}$. The instantaneous temperature profile is decomposed into the average and fluctuating parts,

$$T(z,t) = T_{av}(z) + \delta T(z,t) = T_0 + \alpha z + \delta T(z,t), \tag{6.20}$$

where $\alpha = (T_H - T_C)/h$, Eq. (6.1).

The effect of the temperature gradient on the fluctuations can be studied using the Boussinesq approximation [56, 138]. This is defined through the following two approximations:

1. The density is constant, that is, $\rho = \rho_{av}$.
2. The thermodynamic properties and transport coefficients are constants and independent of the local thermodynamic state.

We have, as the reader may note, used the second approximation throughout the book. Since the density is constant in this approximation, we cannot study density (or sound) waves; hence, we only include the Rayleigh process in the modelling. Also, from the mass balance equation, we have in the steady state

$$\nabla \cdot \rho \mathbf{u} = 0, \tag{6.21}$$

which leads to the instantaneous incompressibility criterion,

$$\rho_{av} \nabla \cdot (\mathbf{u}_{av} + \delta \mathbf{u}) = 0 \text{ implying } \nabla \cdot \delta \mathbf{u} = 0. \tag{6.22}$$

Next, we write up the dynamical equation for the linear momentum fluctuations. From Eqs. (3.25b), (3.48), and (6.22), the dynamics for the streaming velocity fluctuations is

$$\rho_{\mathrm{av}}\frac{\partial \delta \mathbf{u}}{\partial t} = -\beta_V \boldsymbol{\nabla}\delta T + \eta_0 \nabla^2 \delta \mathbf{u} - \boldsymbol{\nabla}\cdot \delta \mathbf{P} \qquad (6.23)$$

to first order in the fluctuations, and using $\rho\,\delta T = \rho\,\delta\varepsilon/c_V$.

Recall, the equation for the kinetic temperature, Eq. (2.91). Since the average flow is zero, we ignore viscous heating and we have

$$\rho_{\mathrm{av}}c_V\left(\frac{\partial T}{\partial t}+\mathbf{u}\cdot\boldsymbol{\nabla}T\right)=-\boldsymbol{\nabla}\cdot\mathbf{J}^\varepsilon. \qquad (6.24)$$

One could easily be tempted to ignore the advection in a first-order approximation since we have zero flow; however, we need to be careful. As $\mathbf{u}=\delta\mathbf{u}$, the advective derivative becomes

$$\begin{aligned}
\mathbf{u}\cdot\boldsymbol{\nabla}T &= \delta\mathbf{u}\cdot\boldsymbol{\nabla}(T_{\mathrm{av}}(z)+\delta T)\\
&= \delta u_z\frac{\partial T_{\mathrm{av}}}{\partial z}+\delta\mathbf{u}\cdot\boldsymbol{\nabla}\delta T\\
&\approx \delta u_z\frac{\partial T_{\mathrm{av}}}{\partial z}=\alpha\delta u_z.
\end{aligned} \qquad (6.25)$$

The advective term is non-zero even to first order in the fluctuations. By application of Fourier's law Eq. (3.22c), the dynamical equation for the temperature fluctuations reads

$$\rho_{\mathrm{av}}c_V\left(\frac{\partial\delta T}{\partial t}+\alpha\delta u_z\right)=\lambda\nabla^2 T-\boldsymbol{\nabla}\cdot\delta\mathbf{J}^\varepsilon. \qquad (6.26)$$

Equations (6.23) and (6.26) are the linearised Boussinesq equations.

Following the approach we used in Chapter 3, we explore the Fourier coefficients. Thus,

$$\frac{\partial\widetilde{\delta\mathbf{u}}}{\partial t}=-\frac{i\beta_V\mathbf{k}}{\rho}\widetilde{\delta T}-\nu_0 k^2\widetilde{\delta\mathbf{u}}-i\mathbf{k}\cdot\widetilde{\delta\mathbf{P}} \qquad (6.27\mathrm{a})$$

$$\frac{\partial\widetilde{\delta T}}{\partial t}=\alpha\widetilde{\delta u_z}-\kappa k^2\widetilde{\delta T}-\frac{i\mathbf{k}}{\rho c_V}\cdot\widetilde{\delta\mathbf{J}}^\varepsilon. \qquad (6.27\mathrm{b})$$

Let us choose a wavevector normal to the z-direction, say $\mathbf{k}=(k_x,0,0)$; then the momentum z-direction decouples from the temperature:

$$\frac{\partial\widetilde{\delta u_z}}{\partial t}=-\nu_0 k_x^2\widetilde{\delta u_z}-i(\mathbf{k}\cdot\widetilde{\delta\mathbf{P}})_z \qquad (6.28\mathrm{a})$$

$$\frac{\partial\widetilde{\delta T}}{\partial t}=-\alpha\widetilde{\delta u_z}-\kappa k_x^2\widetilde{\delta T}-\frac{ik_x}{\rho c_V}\cdot\widetilde{\delta J}_x^\varepsilon. \qquad (6.28\mathrm{b})$$

We can now form the correlation functions C_{uu},C_{uT},C_{Tu}, and C_{TT} obeying the dynamics

$$\frac{\partial}{\partial t}\begin{bmatrix}C_{uu} & C_{uT}\\ C_{Tu} & C_{TT}\end{bmatrix}=\begin{bmatrix}-\nu_0 k_x^2 & 0\\ -\alpha & -\kappa k_x^2\end{bmatrix}\cdot\begin{bmatrix}C_{uu} & C_{uT}\\ C_{Tu} & C_{TT}\end{bmatrix}. \qquad (6.29)$$

Here we have the co-dependent pair $\{C_{uT}, C_{TT}\}$. From the hydrodynamic matrix, we immediately identify two characteristic frequencies (or eigenvalues) for the temperature autocorrelation function C_{TT}, namely,

$$\omega_1 = -\nu_0 k_x^2 \text{ and } \omega_2 = -\kappa k_x^2. \tag{6.30}$$

The relaxation is thus characterised by two exponential processes: (i) a thermal diffusion process and (ii) a viscous process; the latter is not present in equilibrium under the Boussinesq approximation.

If the two processes have different characteristic relaxation times, we can observe both of them in a mechanical spectrum; that is, if the Prandtl number,

$$\Pr = \nu_0/\kappa, \tag{6.31}$$

is not unity. For water, $\Pr \approx 10$ at ambient conditions, whereas for Lennard–Jones-type liquids, $\Pr \approx 1$. The coupling is difficult to explore using direct molecular simulations, since the Boussinesq approximation fails in the presence of large gradients, which are often employed; the reader is referred to Refs. [45, 55] for further treatments of this phenomenon.

6.1.5 The Planar Poiseuille Flow Revisited

Fluid flows generate heat due to the internal fluid friction. For sufficiently small flow rates, the generated heat is conducted away from the fluid through the thermostated walls, and the temperature can be regarded as constant. For nanoscale fluid systems, where the surface-to-volume ratio is relative large, this thermostating is highly effective. Nevertheless, when reaching sufficiently large flow rates in simulations, a temperature gradient can be generated, and here we address this point.

Figure 6.5 shows the relative temperature difference for a simple Lennard–Jones fluid undergoing a planar Poiseuille flow. Symbols are data obtained from molecular dynamics, and T_0 is the wall temperature. The punctured line plots the predictions from the classical theory, or more precisely, where the thermal kinetic energy flux is given by Fourier's law. The classical approach clearly fails. The full line is the result of an extended theory first proposed by Baranyai et al. [12]; before exploring this extended theory, we first revisit the problem from a classical viewpoint.

The geometry is the usual one, and we only need to consider gradients in the z-direction. For steady non-advective flows, but with viscous heating, the balance equation for the kinetic temperature, Eq. (2.91), yields

$$\frac{dJ_z^\varepsilon}{dz} = -P_{xz}\frac{du_x}{dz}. \tag{6.32}$$

Recall Fourier's law,

$$J_z^\varepsilon = -\lambda\frac{dT_{cl}}{dz}, \tag{6.33}$$

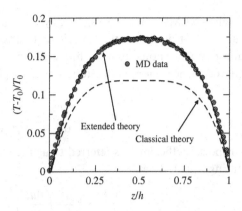

Figure 6.5 Relative temperature difference in a slit-pore for a fluid undergoing a Poiseuille flow. T_0 is the fluid temperature at the walls, and h is 67 σ, σ being the Lennard–Jones length scale parameter. Data are taken from Ref. [196]. The geometry is the same as in Fig. 5.8(a).

where λ is the heat conductivity, and assume this transport coefficient is constant. T_{cl} is used to indicate the classical kinetic temperature. Substitution gives

$$\lambda \frac{\mathrm{d}^2 T_{\text{cl}}}{\mathrm{d}z^2} = P_{xz} \frac{\mathrm{d}u_x}{\mathrm{d}z}. \tag{6.34}$$

We need an expression for the right-hand side. If we also assume that the viscosity is constant, the velocity is given by the standard Poiseuille solution Eq. (5.45), which with zero slip-length reads

$$u_x(z) = \frac{\rho g}{2\eta_0} z(h - z). \tag{6.35}$$

The shear pressure becomes

$$P_{xz}(z) = -\eta_0 \frac{\mathrm{d}u_x}{\mathrm{d}z} = -\frac{\rho g}{2}(h - 2z), \tag{6.36}$$

and the right-hand side of Eq. (6.34) is therefore

$$P_{xz}(z) \frac{\mathrm{d}u_x}{\mathrm{d}z} = -\frac{(\rho g)^2}{\eta_0}(z^2 + h^2/4 - zh). \tag{6.37}$$

Substituting, integrating, and applying the Dirichlet boundary conditions,

$$T_{\text{cl}}(0) = T_{\text{cl}}(h) = T_0, \tag{6.38}$$

yield the solution

$$T_{\text{cl}}(z) = -\frac{(\rho g)^2}{24\lambda \eta_0}\left(2z^3 - 4hz^2 + 3h^2 z + h^3\right)z + T_0. \tag{6.39}$$

Using values for the conductivity λ and viscosity η_0 found from independent simulations [196], we obtain the punctured line given in Fig. 6.5, showing that the classical treatment does not suffice. Note: including the state-point dependency and non-local effects in the heat conductivity will not explain the large deviation.

Through s-NEMD simulations, Baranyai et al. [12] reported an energy flux even in the absence of a temperature gradient; Fourier's law cannot be the whole story. The authors further noted that the flux was proportional to the gradient of the strain rate squared, at least for moderate strain rates [48]. For the present geometry, this means

$$J_z^\varepsilon = -\lambda \frac{dT}{dz} - \xi \frac{d}{dz}\left(\frac{du_x}{dz}\right)^2. \tag{6.40}$$

The transport coefficient ξ is referred to as the strain rate coupling coefficient [200]. Substituting, we arrive at

$$\lambda \frac{d^2 T}{dz^2} = \xi \frac{d^2}{dz^2}\left(\frac{du_x}{dz}\right)^2 + P_{xz}\frac{du_x}{dz}. \tag{6.41}$$

Using Eqs. (6.36) and (5.45), and integrating, we can write the solution in the form

$$T = T_{cl} + T_{ext}, \tag{6.42}$$

where

$$T_{ext} = \frac{\xi(\rho g)^2}{\lambda \eta_0^2} z(h-z) \tag{6.43}$$

is the contribution from the extended theory. Notice that this contribution is small near the wall–fluid boundaries, $z \approx 0$ and $z \approx h$, and becomes maximum at the channel centre. Using the same values for the conductivity and shear viscosity as the classical treatment, but allowing the coupling coefficient to be a fitting parameter, we obtain a very good agreement between the simulation data and the theory.

Both the classical and extended contributions depend on the external force density squared. Therefore, the coupling term cannot be ignored even in the small force limit [200].

6.2 Charged Systems in Confinement

Electrolyte solutions confined between charged surfaces play a very important role in application in both microfluidics and nanofluidics. In the next section, the fundamental theory for how electrolyte distributes in such a system is introduced, and we use the results to explore nanoscale fluid flows and how these are coupled to the distribution of charges.

6.2.1 The Electric Double Layer

Surface charges are not only realised on conductor surfaces, but can also be formed from chemical reactions on the surface. Nano-slit pores fabricated from a silicon wafer is one such example: here the wafer is etched with a strong acid, forming a cavity; and following the etching, the cavity height can be controlled by allowing the pristine silicon

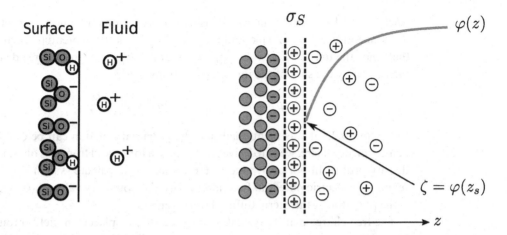

Figure 6.6 (a) Illustration of the dissociation of surface silanol groups in a silica–water system. (b) Illustration of the electric double-layer model.

in the cavity to oxidise, forming silica. After the oxidation process, a silica/silicon lit is placed on top of the cavity, forming a rectangular-shaped pore; see also Fig. 1.1. Impressively, fabrication of pore height down to 2–3 nm can now be controlled [148]. Now, when in contact with de-ionised water, the silica surface is hydroxylated forming silanol groups, $-SiOH$. This group can donate a proton according to the equilibrium reaction scheme [16],

$$- SiOH \rightleftharpoons -SiO^- + H^+. \tag{6.44}$$

The proton can enter reactions occurring in the fluid, like the auto-dissociation mechanism of water, and is thus no longer located near the surface. This means that the surface becomes charged, while the entire system still features charge neutrality; see the illustration in Fig. 6.6(a).

If we consider a solution composed of cations and anions (an electrolyte) with charges $\pm q$, respectively, and focus on the wall–fluid interface, we can suggest a conceptual model like the one illustrated in Fig. 6.6(b). Near the charged surface, counter ions pack, and for sufficiently small thermal energies (strictly, at zero kinetic energy) this packing forms an immobile layer of the counter ions that screens the surface charge. The immobile layer is denoted the Stern layer and has a width, σ_S, of approximately one counter ion diameter. Further away from the wall, there exists a diffuse layer (also known as the Gouy–Chapman layer) wherein the ions are mobile. These two different layers form the basic picture behind the electric double layer (EDL) model.

We now choose the coordinate system such that $z = 0$ at the Stern–diffuse layer boundary. The ion distribution in the diffuse layer can be described through the Boltzmann distribution introduced in Section 5.2.1, that is,

$$n_+ = n_0 e^{-q\varphi_q/k_B T} \text{ and } n_- = n_0 e^{q\varphi_q/k_B T}, \tag{6.45}$$

where n_+ and n_- are the number densities of the cation and anion, respectively, n_0 is the concentration in the bulk, and $\varphi_q = \varphi_q(z)$ is the electric potential function. Notice that since the unit of φ_q is volt, $q\varphi_q$ has unit of energy. The free charge density, ρ_f, is then given by the sum of the ion concentrations, that is,

$$\rho_f = q(n_+ - n_-) = -2qn_0 \sinh\left(\frac{q\varphi_q}{k_B T}\right). \tag{6.46}$$

In Section 5.2.1 it was mentioned that the potential function can be evaluated from density profile. Here, we will derive an expression for φ_q and from this obtain the charge density profile. Importantly, due to the choice of coordinate system, Fig. 6.6(b), we require the electric potential function to have the boundary value $\varphi_q(0) = \zeta$, where ζ is the potential at the Stern–diffuse layer interface.

The free-charge density is related to the electric displacement field through Gauss' law,

$$\nabla \cdot \mathbf{D} = \rho_f. \tag{6.47}$$

In the simplest case, the system is linear and homogeneous. We then assume that the polarisation is given by the local electric field, $\mathbf{E}$,

$$\mathbf{P} = \varepsilon_0(\varepsilon_r - 1)\mathbf{E}. \tag{6.48}$$

The displacement field is in this linear regime

$$\mathbf{D} = \varepsilon_0\mathbf{E} + \mathbf{P} = \varepsilon\mathbf{E}, \tag{6.49}$$

where $\varepsilon = \varepsilon_0\varepsilon_r$ is the system permittivity. Gauss' law can therefore be expressed as a Poisson equation for the electric potential function, $\nabla\varphi_q = -\mathbf{E}$; substitution gives

$$\nabla^2\varphi_q = -\frac{\rho_f}{\varepsilon}. \tag{6.50}$$

For the current geometry, this means that

$$\frac{d^2\varphi_q}{dz^2} = \frac{2qn_0}{\varepsilon}\sinh\left(\frac{q\varphi_q}{k_B T}\right), \tag{6.51}$$

which is the one-dimensional Poisson–Boltzmann equation.

The one-dimensional Poisson–Boltzmann equation has been solved for the special case where $\varphi_q(z) \to 0$ as $z \to \infty$, which corresponds to studying the electric potential at a single wall such that $n_+ = n_- = n_0$ in the bulk. This gives the so-called Gouy–Chapman solution,

$$\varphi_q(z) = \frac{4k_B T}{q}\tanh^{-1}\left[\tanh\left(\frac{q\zeta}{4k_B T}\right)e^{-z/\lambda_D}\right], \tag{6.52}$$

where λ_D is the Debye length; more on this in what follows.

Rather than showing how to arrive at this solution,[1] we will simplify the problem, retaining the important point. In the Debye–Hückel limit, the thermal energy is large compared to the electric energy:

$$q\varphi_q \ll k_B T. \tag{6.53}$$

---

[1] Bruus [38] details this in a student exercise.

Notice that this relation cannot be true in the immobile Stern layer. In the Debye–Hückel limit, $\sinh(q\varphi_q/k_BT) \approx q\varphi_q/k_BT$, and we have the linear problem

$$\frac{d^2\varphi_q}{dz^2} = \frac{2qn_0}{\varepsilon k_BT}\varphi_q. \tag{6.54}$$

We solve this using the boundary values

$$\varphi_q(0) = \zeta \text{ and } \lim_{z\to\infty}\varphi_q(z) = 0, \tag{6.55}$$

giving

$$\varphi_q(z) = \zeta\, e^{-z/\lambda_D}, \tag{6.56}$$

where the characteristic length scale,

$$\lambda_D = \sqrt{\frac{\varepsilon k_BT}{2q^2 n_0}}, \tag{6.57}$$

is the Debye length or the Debye layer. As an example from Ref. [38], for a monovalent, $q = 1.6 \times 10^{-19}$ C, binary electrolyte solution with concentration $c = n_0/N_A = 1.0\,\text{mM} = 1.0\,\text{mol m}^{-3}$ at $T = 298$ K, we can assume that the permittivity is that of water, $\varepsilon = \varepsilon_r\varepsilon_0 = 78\varepsilon_0$, giving a Debye length of approximately 9.6 nm. λ_D can be thought of as a characteristic screening length of the surface charge; hence, it is a nanoscale phenomenon and is particularly important for nanoscale fluid systems, as we have the relation

$$\lambda_D \approx h. \tag{6.58}$$

From Eq. (6.56), the free-charge density is found directly by differentiation,

$$\rho_f(z) = -\frac{\varepsilon\zeta}{\lambda_D^2}e^{-z/\lambda_D}. \tag{6.59}$$

As just stated, the effect of the mobile charges is that of screening. We have from global charge neutrality that

$$\lim_{z\to\infty} A\int_{-\sigma_S}^{z}\rho_f(z')\,dz' + \Sigma_W = 0, \tag{6.60}$$

where Σ_W is the total wall charge, and A is the wall surface area. The screening charge is given by the first term, which in the limit $\sigma_S \ll \lambda_D$ is

$$q_{\text{scr}}(z) = \frac{A\varepsilon\zeta}{\lambda_D}(1 - e^{-z/\lambda_D}). \tag{6.61}$$

For nanoscale slit-pores, this means that the screening from each wall is not complete, as h can be the same order of magnitude as the Debye length. This leads to a so-called Debye-layer overlap, and this affects the flow profile of electrolyte mixtures considerably, as we will see in what follows.

Joly et al. [120, 121] did a study of a model electrolyte solution, where the solvent molecules and ions were simple Lennard–Jones-type particles. From their ion density

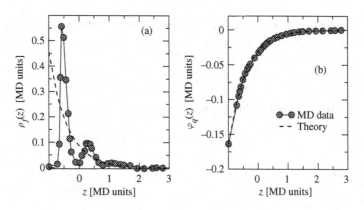

Figure 6.7 (a) Charge density profile. (b) Electric potential profile. In both (a) and (b) the circle symbols connected with lines represent results derived from direct-NEMD simulation by Joly et al. [120]. The dashed lines represent the Poisson–Boltzmann theory in the Debye–Hückel limit, using $\lambda_D = 0.6$ and $\zeta = -0.03$; values are given in molecular dynamics units.

profiles, the charge density can be calculated directly, Eq. (6.46), and the result is plotted in Fig. 6.7(a). Notice that the Stern layer is also shown here and is approximately unity, hence, the abscissa starts at negative one. The exponential prediction from Eq. (6.59) is also plotted. The charge density clearly features a layering-type profile, and the simple exponential function predicted by the theory (under the Debye–Hückel assumption that $q\zeta \ll k_B T$) will, of course, not capture the ionic layering.

Once the charge-density profile is available, the corresponding electric potential function can be evaluated from the Poisson equation, that is, by integration of the charge density profile. The constants of integration are found by demanding global charge neutrality, Eq. (6.60), and $\lim_{z \to \infty} \varphi_q = 0$. In Fig. 6.7(b) the electric potential is plotted for both the numerical integration of the molecular dynamics data and for Eq. (6.56). One sees that the integration smears out the effect of the layering in the charge density, and the simple Poisson–Boltzmann/Debye–Hückel theory compares very well with the detailed molecular simulations even if the Debye–Hückel limit is not strictly met.

The unsatisfactory theoretical prediction for the charge density needs further comment. In Section 5.2.1 we saw the existence of fluid layering in the wall–fluid interface region. The positions of the ions must be coupled to the solvent layering, and we can model this coupling through the potential function $\varphi = \varphi_c + q\varphi_q$, that is, we assume a potential given by a sum of the solvent potential function and the electrical potential function. In the limit of low electrolyte concentration (weak electrolyte), we further assume that the solvent density $\rho(z)$ is independent of the ion layering; hence, from Eq. (5.25),

$$\rho(z) = \rho_0\, e^{-\frac{\varphi_c}{k_B T}}. \tag{6.62}$$

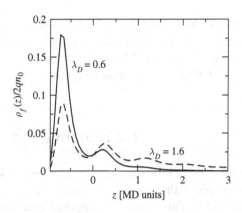

Figure 6.8 Charge-density profiles for different Debye lengths calculated from Eq. (6.65). The Debye length is given in dimensionless MD units.

The ion number density distributions can then be rewritten as [120, 121]

$$n_+ = n_0 e^{-\frac{q\varphi_q + \varphi_c}{k_B T}} = \frac{n_0}{\rho_0}\rho(z)e^{-q\varphi_q/k_B T} \tag{6.63}$$

and

$$n_- = n_0 e^{\frac{q\varphi_q - \varphi_c}{k_B T}} = \frac{n_0}{\rho_0}\rho(z)e^{q\varphi_q/k_B T}. \tag{6.64}$$

If we let φ_q be given by Eq. (6.56), we have for the charge density [121]

$$\rho_f(z) = -\frac{2qn_0}{\rho_0}\rho(z)\sinh\left(\frac{q\zeta}{k_B T}e^{-z/\lambda_D}\right). \tag{6.65}$$

In Fig. 6.8 the charge density based on Eq. (6.65) is plotted for two different Debye lengths, λ_D. The solvent density profile used for the calculation is similar to Fig. 5.2(b).

We see that for small Debye lengths the system effectively screens the wall charge and the charge density layering quickly decays. On the other hand, for large Debye lengths the charge screening is reduced and a clear layering is seen many molecular diameters away from the wall. Thus, for fixed system permittivity and temperature, this layering will be pronounced in the dilute electrolyte regime, and the structure can span several nanometres into the slit-pore even for simple electrolyte solutions.

To form a self-contained closed theory, we must provide an expression of the solvent density profile $\rho(z)$; however, such an endeavour is outside the scope of this text; see advanced theories specifically for the EDL [17, 166].

On purpose, the term ζ-*potential* has not been used for the boundary condition $\varphi_q(0) = \zeta$. In the literature the ζ-potential has slightly different definitions; for example, Bruus [38] defines it to be the electric potential at the wall surface, and Karniadakis et al. [127] as the electric potential at the Stern layer and diffuse layer interface, $z = z_S$. The latter is a special case of the more general definition [189] that the ζ-potential is the electric potential at the so-called slipping plane where the electrolyte solution will flow by application of shear stress; for many systems, the slipping plane resides inside the diffuse layer and not at the interface z_S. See also Refs. [120, 121] for a discussion of this. Nevertheless, the mathematical treatment is, of course, independent of the exact choice.

6.2.2 Counter-ion Systems

Before we explore how the electrolyte nanoscale layering affects the flow properties, it is worth presenting another classical charged system, which is encountered in the literature [117]. This system is composed of a charged wall; but rather than an electrolyte solution consisting of wall charge co-ions and counter-ions, the solution only contains counter-ions. We can again use the Boltzmann distribution function; thus, if the surface is negatively charged, the cation concentration is

$$n_+ = n_0(e^{-q\varphi_q/k_BT} - 1). \tag{6.66}$$

In contrast to the electrolyte solution, the cation concentration will here approach zero as $z \to \infty$ due to the screening and charge neutrality requirement. Thus, n_0 is here some reference concentration. The Poisson–Boltzmann equation reads

$$\frac{d^2\varphi_q}{dz^2} = -\frac{qn_0}{\varepsilon}(e^{-q\varphi_q/k_BT} - 1), \tag{6.67}$$

which in the Debye–Hückel limit becomes

$$\frac{d^2\varphi_q}{dz^2} = \frac{q^2n_0}{\varepsilon k_BT}\varphi_q. \tag{6.68}$$

This is solved using

$$\varphi_q(0) = \zeta \text{ and } \lim_{z\to\infty}\varphi_q'(z) = 0, \tag{6.69}$$

the latter being the charge neutrality condition. This yields

$$\varphi_q(z) = \zeta e^{-z/(\sqrt{2}\lambda_D)}. \tag{6.70}$$

The corresponding charge density follows the same simple exponential form as for the electrolytes, namely,

$$\rho_f(z) = -\frac{\varepsilon\zeta}{2\lambda_D^2}e^{-z/(\sqrt{2}\lambda_D)}. \tag{6.71}$$

Molecular dynamics simulations of counter-ion systems [121] reveal the same layering phenomenon as seen in the electrolyte system, Fig. 6.7; hence, to model the charge profile more accurately, we can include the solvent layering as we did previously.

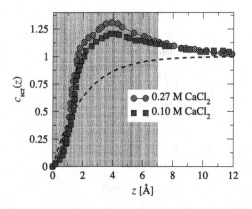

Figure 6.9 Screening function c_{scr} for different $CaCl_2$ electrolyte solutions (symbols connected with lines) confined between charged silica walls. The punctured line shows the functional form of the Poisson–Boltzmann result in the Debye–Hückel limit. $z = 0$ is here the first non-zero density point in the cation concentration; the approximate Stern layer is indicated by the grey region. Data are taken from Ref. [59].

6.2.3 Over-screening and the Ferroelectric Effect

The charge layering in the wall–fluid interface can give rise to an over-screening. This phenomenon can be quantified by a screening function [33]; for a negatively charged surface, we write this as

$$c_{scr}(z) = \frac{q_{scr}(z)}{|\Sigma_W|}, \qquad (6.72)$$

such that c_{scr} is unity for perfect screening, zero for no screening, and above one shows an over-screening of the wall charge. Figure 6.9 shows the screening function for $CaCl_2$ in water confined between two charged amorphous silica walls [59]. For comparison the standard Poisson–Boltzmann result in the Debye–Hückel limit is also shown. Clearly, in the Stern layer the counter-ions over-screen the wall charge. The over-screening is naturally not accounted for in the Debye–Hückel theory.

The over-screening phenomenon is not only observed for mobile charges. Fig. 6.10(a) shows the charge density profile obtained from molecular dynamics for water confined in a nano-slit-pore. The walls are non-charged; hence, there exists a local "over-screening" in the wall–fluid interface. From the charge distribution we can calculate the electric field shown in Fig. 6.10(b). This induced local electric field causes a spontaneous polarisation of the dielectric medium and is a fingerprint of the so-called ferroelectric effect.

The ferroelectric effect emerges due to molecular alignment and packing near the wall and is thus geometrically induced. It will change the dielectric response of the system to an external electric field, because in the wall–fluid interface the molecular dipole rotation is suppressed in the out-of-plane rotation with respect to the wall. Thus, the dipole orientation with an external field applied normal to the walls will be lowered compared to the bulk, and this results in a reduced dielectric permittivity. This reduced

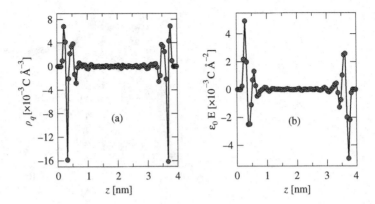

Figure 6.10 Water in a nano-slit-pore where the walls are neutral. (a) Charge density profile. (b) Electric field profile. Courtesy of S. Varghese.

permittivity was presented in Fig. 1.4. It has been conjectured that the reduced out-of-plane rotation is due to the increasing hydrogen-bond density in the wall–fluid interface [210]; again, an effect due to geometrical constraint.

6.2.4 Electro-osmosis in Nano-confinement

An electro-osmotic flow, abbreviated EOF, is a flow generated by application of an electric field parallel to the charged walls such that the fluid flows with respect to these surfaces. This is illustrated in Fig. 6.11. We assume that the screening effect is sufficiently small, also, at the electrodes. The electric field then exerts a constant body-type force on each ion, and as the ions migrate, they will move the entire fluid due to momentum transferal.

For the geometry shown in Fig. 6.11 and for sufficiently low flow velocities, the Navier–Stokes equation is in the same form as Eq. (5.33); however, here the force density is given by $\rho_f E_x$, where E_x is the local field, which we assume is constant. The charge density ρ_f is, on the other hand, a function of z, as we have shown; hence, the local force is not constant, and we must expect that the flow is very different from the planar Poiseuille flow. We have

$$\rho \frac{\partial u_x}{\partial t} = \rho_f E_x + \eta_0 \frac{\partial^2 u_x}{\partial z^2}. \tag{6.73}$$

To proceed, we must find ρ_f, and to this end we use the equilibrium assumption, that the charge density is unaffected by the flow. In the Debye–Hückel limit we solve the linearised Poisson–Boltzmann equation, Eq. (6.54), with the boundary conditions

$$\varphi_q(0) = \varphi_q(h) = \zeta. \tag{6.74}$$

This gives

$$\varphi_q(z) = \zeta \phi(z), \tag{6.75}$$

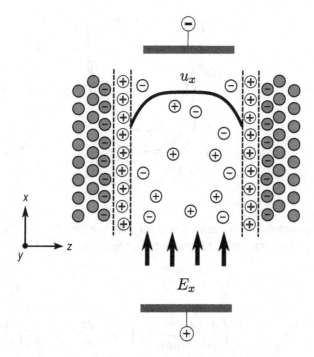

Figure 6.11   An electro-osmotic flow (EOF) generated by an external electric field parallel to the x-axis. The resulting velocity, $u_x = u_x(z)$, also is illustrated.

where, out of convenience, we have introduced the function

$$\phi(z) = e^{-z/\lambda_D} + \frac{1 - e^{-h/\lambda_D}}{\sinh(h/\lambda_D)} \sinh(z/\lambda_D). \tag{6.76}$$

Differentiating twice, the corresponding charge density is found to be

$$\rho_f = -\frac{\varepsilon \zeta}{\lambda_D^2} \phi(z). \tag{6.77}$$

Substituting into Eq. (6.73) and applying the steady-state conditions, that is, $u_x = u_x(z)$, gives

$$\frac{d^2 u_x}{dz^2} = \frac{\varepsilon \zeta E_x}{\eta_0 \lambda_D^2} \phi(z), \tag{6.78}$$

which we solve with slip-boundary conditions at the Stern–diffuse layer interface,

$$L_s \left. \frac{du_x}{dz} \right|_{z=0} = u_x(0) \text{ and } L_s \left. \frac{du_x}{dz} \right|_{z=h} = -u_x(h). \tag{6.79}$$

From integration, it is seen that the general solution is in the form

$$u_x(z) = \frac{\varepsilon E_x \zeta}{\eta_0} \phi(z) + K_1 z + K_2, \tag{6.80}$$

where K_1 and K_2 are integration constants. Using the properties of ϕ,

$$\phi(0) = \phi(h) = 1 \tag{6.81}$$

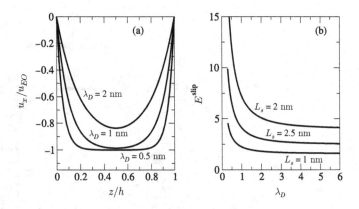

Figure 6.12   (a) Normalized electro-osmotic flow velocity profile for different Debye lengths. $L_s = 0$. (b) The flow enhancement coefficient as a function of λ_D and for different slip lengths. h is 10 nm in both (a) and (b).

and

$$\frac{d\phi}{dz}\bigg|_{z=0} = \frac{1}{\lambda_D}\left(\frac{1-e^{-h/\lambda_D}}{\sinh(h/\lambda_D)}-1\right) \text{ and } \frac{d\phi}{dz}\bigg|_{z=h} = \frac{1}{\lambda_D}\left(\frac{1-e^{-h/\lambda_D}}{\tanh(h/\lambda_D)}-e^{-h/\lambda_D}\right), \quad (6.82)$$

it can be seen that K_1 is zero for finite channel heights. The solution is written in terms of the electro-osmotic or Helmholtz–Smoluchowski velocity,

$$u_{EO} = \frac{\varepsilon E_x \zeta}{\eta_0}, \quad (6.83)$$

giving

$$u_x(z) = u_{EO}\left[\phi(z) - 1 + \frac{L_s}{\lambda_D}\left(\frac{1-e^{-h/\lambda_D}}{\sinh(h/\lambda_D)}-1\right)\right]. \quad (6.84)$$

Clearly, if $L_s = 0$ the streaming velocity is simply $u_x = u_{EO}(\phi - 1)$. In this case and in the limit of small Debye lengths, the velocity approaches $-u_{EO}$ in the interior of the channel. This is also clear from Fig. 6.12(a) which shows the velocity profile in the slit-pore for different Debye lengths; for small λ_D, the flow features a plug-type profile with a magnitude of the electro-osmotic velocity.

In the general case, $L_s \neq 0$, the magnitude of the velocity exceeds the Helmholtz–Smoluchowski velocity, and the flow rate is increased significantly. This enhanced can be quantified by the enhancement coefficient already defined in Eq. (5.51), that is, by evaluating the volumetric flows rate with and without slippage. The actual analytical expression is left to the reader in *Further Explorations*. The result is plotted in Fig. 6.12(b), showing that the flow rate is significantly increased in the regime of small Debye lengths and large slip length. This combined effect is thus very important to consider when studying electric currents through nanoscale confinement.

6.3 Reactive Fluids

By letting a reactive fluid diffuse into a nanochannel, the channel walls can be coated with a specific product resulting from the reactions between the reactive fluid and the wall atoms. By repeating such a procedure with different reactive fluids, the channel walls can be designed with patterns, giving unique functionality [128, 164]. Porous catalyst pellets with nanoscale reaction channels can speed up the product output rate considerably, and these systems are therefore of great industrial importance [192].

In this last section we investigate some classical reaction models and set them in the context of nanoscale fluid systems. The underlying mathematical model we apply here is the reaction-diffusion equation. This is based on the single-particle dynamical equation, Eq. (3.90), wherein we allow for a production term that accounts for the reactions. We naturally denote this *the reaction term*. For molecules of type A we get, in absence of advection and in terms of the number density n_A,

$$\frac{\partial n_A}{\partial t} = \sigma_A - \mathbf{\nabla} \cdot \mathbf{J}^A. \tag{6.85}$$

The reaction term is modelled from the law of mass action, which states that the reaction rate, r, of a reaction is proportional to the product of the reactants' concentrations to their stoichiometric powers. For example, if n molecules of type A react with m molecules of B, giving product P,

$$n\mathrm{A} + m\mathrm{B} \xrightarrow{k} \mathrm{P}, \tag{6.86}$$

the reaction rate is

$$r = k\, n_A^n\, n_B^m. \tag{6.87}$$

The constant k is the reaction rate constant. The law of mass action can be derived by assuming that the molecules undergo independent collision events, and it is therefore valid only in the limit of infinite dilution. Nevertheless, due to the very high collision frequency in liquids, it is used successfully also for non-ideal mixtures.

The single-particle flux, $\mathbf{J}^A$, follows Fick's law, introduced in Section 3.2:

$$\mathbf{J}^A = -D_A \mathbf{\nabla} n_A. \tag{6.88}$$

We will let D_A be constant; that is, it does not depend on time, spatial coordinate, and the reactive mixture's composition. This is also strictly true only in the infinite dilute limit. In our exploration we will have only one independent component, A, and Eq. (6.85) is written in the final form,

$$\frac{\partial n_A}{\partial t} = \sigma_A(n_A) + D_A \mathbf{\nabla}^2 n_A. \tag{6.89}$$

Of course, σ_A must be specified for each reactive system. This is the reaction-diffusion equation that we base our modelling efforts on. In the next section we will investigate one important effect of fluctuations that are due to the reactions, and not, as we have been focusing on thus far, fluctuations from the particle flux. We therefore do not need to include the fluctuating part of the particle flux $\delta \mathbf{J}^A$.

6.3.1 Homogeneous Reactions: Fluctuation Death

First, it is enlightening to consider a simple homogeneous system where we ignore diffusion. The system is composed of two reactants, namely A and B, which are submerged in a solvent S. The concentration of B is held fixed; this can be achieved by continuously adding a solution of B into the reaction chamber. If n_A denotes the concentration of A, and b_0 the fixed concentration of B, the reaction mechanism we explore reads

$$A + B \xrightarrow{k_1} 2A, \qquad r_1 = k_1 n_A b_0,$$

$$2A \xrightarrow{k_2} 2P, \qquad r_2 = k_2 n_A^2,$$

where P is the product, r_1 and r_2 are the reaction rates, and k_1 and k_2 the corresponding rate constants. The rate of change for A is then written in terms of the number density, $n_A = n_A(t)$,

$$\frac{dn_A}{dt} = \sigma_A = r_1 - 2r_2 = n_A(k_1 b_0 - 2k_2 n_A). \tag{6.90}$$

For this system we have two steady states, denoted n_1 and n_2, respectively:

$$\frac{dn_A}{dt} = 0 \quad \Rightarrow \quad n_1 = 0 \text{ and } n_2 = \frac{k_1 b_0}{2k_2}. \tag{6.91}$$

The point n_1 is unstable and n_2 is stable; this can be seen by plotting the right-hand side of Eq. (6.90) as a function of n_A, Fig. 6.13. If we start with $0 < n_A < n_2$, the rate of change is positive and n_A increases converging to n_2; if we start with $n_A > n_2$, it decreases, again converging to n_2. Thus, if we initialise the system as $n_A \neq 0$, we have $n_A \to n_2$ as $t \to \infty$. If at any point in time $n_A = 0$, the rate of production of A stops and n_A is zero indefinitely; the system has ended in the unstable state.

In nanoscale systems, the number of reactive molecules can be quite small and feature fluctuations in the number of molecules as the reactions take place. In constant-volume geometries, this means that the concentrations fluctuate. The effect of this can be investigated, and molecular dynamics again provides an approach to this. One can simulate a simple Lennard–Jones fluid and label the particle A, B, or P. We ensure that the number of P particles is much larger than the number of A and B particles; thus, they act as a solvent. Whenever an A and a B particle, or two A particles, are sufficiently close, the labels change with some probability in accordance with the reaction scheme. To keep the number of B particles fixed, we pick randomly a P particle and convert it to a B particle, whenever one B is consumed in a reaction. Homogeneity is ensured by frequently interchanging labels. It has been shown that this method yields the same system statistical properties as the Master equation [97] and follows the law of mass action in the dilute case, as expected. The molecular dynamics method is significantly slower than many other numerical methods, but offers a straightforward

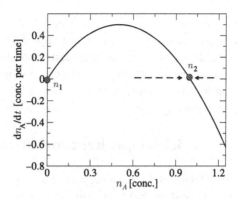

Figure 6.13 Rate of change, dn_A/dt, as a function of n_A. $k_1 b_0/2k_2 = 1$ in some arbitrary concentration unit. The arrows indicate the increase or decrease of n_A, depending on the initial value.

extension to study non-homogeneous systems, fast reactions, and so on. We will see one such application later.

Figure 6.14 shows the results from two molecular dynamics simulations of the reaction. In one simulation the number of A particles fluctuates around the stable steady state, but in the other simulation the fluctuations force the system into the unstable steady state; and once it is in this state, the reactions stop. In fact, the system will end up in this unstable state for sufficiently large times; this waiting time increases dramatically as the system size increases, since the fluctuation amplitude is on the order of $1/\sqrt{N_{\text{reac}}}$, where N_{reac} is the number of reactive molecules.

Again, a first modelling approach that comes to mind is a Langevin-type equation of the form

$$\frac{dn_A}{dt} = n_A(k_1 b_0 - 2k_2 n_A) + \varepsilon F(t), \tag{6.92}$$

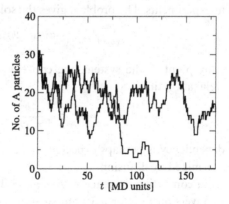

Figure 6.14 Molecular dynamics simulations of the reaction mechanism $A + B \rightarrow 2A \rightarrow 2P$. Initially, the system contains 30 A particles, and the number of B particles is fixed to 30 throughout the simulation.

where F is a noise term with zero mean and amplitude $\varepsilon \propto 1/\sqrt{N_{\text{reac}}}$. While this approach appears straightforward and appealingly simple, it has been shown that it does not yield the correct probability distribution function of the reactants even if the noise amplitude follows the fluctuation-dissipation theorem [13]. The theory for chemical fluctuations is therefore outside the scope of the book, and it suffices to point out here the awareness of unexpected dynamics due to these fluctuations.

6.3.2 Simple Reactions in Nano-geometries

A simple model for surface coating is the Langmuir isothermal adsorption model. Here a reactant, molecule A, binds to a surface site, S, forming a coating product, P; see Fig. 6.15(a). The binding kinetics can be described from the chemical equilibrium

$$A + S \underset{k_2}{\overset{k_1}{\rightleftharpoons}} P, \tag{6.93}$$

where k_1 and k_2 are the reaction rate constants. In the situation where the concentration of binding sites is large compared to the concentration of A, and $k_1 \gg k_2$, we can assume that all A molecules are adsorbed as they diffuse to the wall surface. We will assume that fluctuations play no critical role for the dynamics.

Let n_A denote the number density of A and consider the geometry shown in Fig. 6.15(a); we then have the Dirichlet boundary $n_A = 0$ at $z = 0$, also known as reactive boundaries for chemical reactive systems. At the wall located at $z = h$, we have a non-reactive wall and apply the Neumann boundary condition. The problem is then reduced to that of a simple diffusion equation problem with mixed boundary conditions,

$$\frac{\partial n_A}{\partial t} = D_A \frac{\partial^2 n_A}{\partial z^2} \text{ with } n_A(0,t) = 0 \text{ and } \left. \frac{\partial n_A(z,t)}{\partial z} \right|_{z=h} = 0. \tag{6.94}$$

Without compromising the point in what follows, we let the initial concentration be very convenient, namely, $n_A(z,0) = a_0 \sin(\pi z/2h)$ and do not worry about how this is realised in experiments. The problem gives the solution

$$n_A(z,t) = a_0 e^{-\pi^2 D_A t/(2h)^2} \sin\left(\frac{\pi}{2h}z\right). \tag{6.95}$$

Thus, at any point in the system, the concentration decays with the characteristic diffusive relaxation time,

$$\tau = \frac{4h^2}{\pi^2 D_A}. \tag{6.96}$$

This h^2 dependency is perhaps expected, since the adsorption at the surface is completely determined by diffusion.

The rate of converting the reactant A into product P is then proportional to $D_A h^{-2}$; thus, the conversion rate for a 10 nm slit-pore is around 10^{12} times larger than the conversion rate for a 1 cm channel. This very promising optimisation is, however, limited by the fact that for nanoscale confinements the reactants must first diffuse into

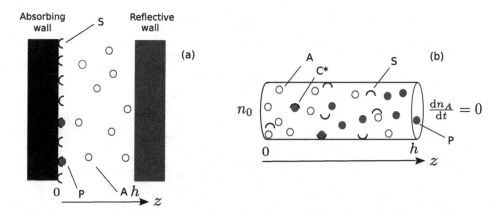

Figure 6.15 (a) Langmuir adsorption. (b) Conversion of reactant A into product P along a pore and in the presence of a catalysis.

the chamber or the chamber can be filled through capillary action; in either case, this reduces the overall effectiveness of such a device.

Next, we will explore a standard problem in catalysis; see, for example, Thomas and McGaughey [192]. Here the reactant A diffuses into a pellet pore and undergoes a reaction in presence of a catalyst, S; see Fig. 6.15(b). Notice that the catalyst is uniformly distributed in the pore and not attached to the pore surface. The overall reaction mechanism is written as

$$A + S \underset{k_2}{\overset{k_1}{\rightleftharpoons}} C^* \overset{k_3}{\to} S + P, \tag{6.97}$$

that is, A and S form a complex C^* which can result in either the product and the catalyst itself or simply reverse to A and S. This is the Michaelis–Menten enzyme mechanism.

From the reaction mechanism we see that the dynamics of A, S, and C^* define the problem. We here focus on the special case where the concentration of C^* is small, and where the concentration of the catalyst is constant with respect to both time and space. This leaves only the concentration of A relevant. Now, the reaction is not localised to the walls, but occurs at any point inside the pore, and the local rate of change must include contributions from diffusion and reactions in which A is involved. Furthermore, it also means that the concentration of A is independent of the radial coordinate r, and we have that $n_A = n_A(z,t)$. We therefore write the rate of change as the reaction-diffusion equation,

$$\frac{\partial n_A}{\partial t} = -k n_A + D_A \frac{\partial^2 n_A}{\partial z^2}, \tag{6.98}$$

where $k = k_1 s_0$ is a pseudo–first-order rate constant, s_0 being the constant catalyst concentration. We look for the steady-state solution, $n_A = n_A(z)$,

$$\frac{d^2 n_A}{dz^2} - \frac{k}{D_A} n_A = 0 \tag{6.99}$$

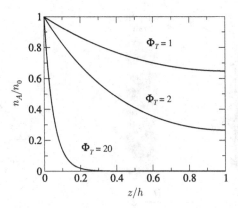

Figure 6.16 The concentration profiles for different Thiele modulus.

with

$$n_A(0) = n_0 \text{ and } \left.\frac{dn_A}{dz}\right|_{z=h} = 0. \qquad (6.100)$$

The first boundary basically expresses the inlet reactant concentration, and the second boundary that the concentration of A for $z > h$ is the same as the outlet concentration. The boundary value problem solves to

$$n_A(z) = n_0 \left[\frac{e^{-\Phi_T}}{\cosh(\Phi_T)} \sinh(\Phi_T z/h) + e^{-\Phi_T z/h}\right], \qquad (6.101)$$

where $\Phi_T = h(k/D_A)^{1/2}$ is the Thiele modulus.

In Fig. 6.16 the concentration profile for A is plotted for different Thiele modulus. Naturally, from a performance point of view we want the concentration of A to be zero at the tube outlet; this is achieved only in the limit of large Φ_T. If $h = 100$ nm and we wish a performance corresponding to $\Phi_T = 10$, the rate constant must be on the order of 10^{12}m^{-2} D_A; that is, the reaction must be extremely fast compared to the diffusion, and this type of homogeneous nanoscale catalysis is relevant mostly for fast reactions. In Section 5.2.3 we saw that for extreme confinement the diffusion coefficient is reduced; however, this reduction is not sufficient to change the conclusion from the analysis above.

6.3.3 Nanoscale Chemical Wave Fronts

Autocatalytic reactions form an important class of chemical reactions. One example is the iodate–arsenite reaction that follows the overall reaction mechanism [84, 89],

$$IO_3^- + 5I^- + 3H_3AsO_3 \rightarrow 6I^- + 3H_3AsO_4. \qquad (6.102)$$

Notice that iodide I$^-$ acts as both reactant and product, and that one iodide is produced per reaction; hence, this is a iodide autocatalysis reaction.

In a homogeneous mixture of iodate IO$_3^-$ and arsenite AsO$_3^{3-}$ a chemical wave front develops by addition of iodide at some point. Ahead of the front, we have the homogeneous mixture of iodate and arsenite, and behind the front, a mixture of iodide and

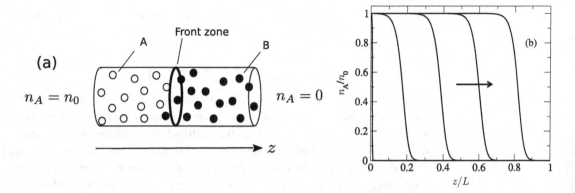

Figure 6.17 (a) Schematic illustration of the Fisher–Kolmogorov front. (b) Concentration profile for A at equidistant times; these are found from the numerical solution to Eq. (6.106).

arsenite. The realisation of the front is due to the presence of both reaction and diffusion processes. The actual front speed depends on the concentrations; for iodate concentrations on the order of 10^{-3}M and arsenite in excess, the front speed is around 0.02 mm s^{-1}.

Reaction (6.102) is the overall reaction scheme. The detailed reaction kinetics for the iodate–arsenite is quite complicated, and it is, perhaps, more enlightening to explore a simple model autocatalytic reaction. Let n_A denote the density of A, and n_B the density of B; then

$$A + B \xrightarrow{k} 2A \qquad \text{with} \qquad r = k n_A n_B. \tag{6.103}$$

For a front to develop, the system is initialised such that we start with only A at the inlet, $z = 0$, and B everywhere else:

$$n_A(z,0) = n_0(1 - H(z)) \quad \text{and} \quad n_B(z,0) = n_0 H(z), \tag{6.104}$$

where H is the Heaviside step function,

$$H(z) = \begin{cases} 0 & \text{if } z \leq 0 \\ 1 & \text{if } z > 0 \end{cases}. \tag{6.105}$$

As for the iodate–arsenite reaction, a chemical wavefront develops due to diffusion and the autocatalytic nature of the reaction. Rigorously proving that such a propagating wave exists (and is unique) has not yet been done in general [162]. However, under the fundamental assumption that the propagating wave does exist and has a constant speed, c, it is possible to show that this speed must be equal to or larger than some minimum speed $c_{\min}$. This speed depends on the physico-chemical constants. From numerical solutions of the reaction-diffusion equation, see Fig. 6.17, the minimum speed has been confirmed; in fact, $c_{\min}$ is the speed of the corresponding stable propagating wavefront solution when the initial conditions are given by Eq. (6.104).

First, we let A and B have the same diffusion coefficient, D_A. Then $n_A + n_B = n_0$ at any point, and the reaction rate is written as $r = k n_A(n_0 - n_A)$. For convenience we introduce a dimensionless density, $u = n_A/n_0$, which fulfils $0 \le u \le 1$, and we have the reaction-diffusion equation for u,

$$\frac{\partial u}{\partial t} = k^* u(1-u) + D_A \frac{\partial^2 u}{\partial z^2}. \tag{6.106}$$

We have used $k^* = kn_0$; we will omit the asterisk from here on. This is known as the Fisher–Kolmogorov equation and has been used to study gene population spread [73] and much more [162]; here it serves as the prototype for a chemical wavefront.

We *assume* that the front exists and has a constant propagation speed. Notice that, in general, this can be true only in the large time regime where transient behaviour has fully decayed. We can follow the wavefront in the frame moving with speed c by a coordinate transformation $x = z - ct$, and we write $u(z,t) = u(x)$. By the chain rule,

$$\frac{\partial u}{\partial t} = -c \frac{du}{dx} \quad \text{and} \quad \frac{\partial^2 u}{\partial z^2} = \frac{d^2 u}{dx^2}. \tag{6.107}$$

Substitution into Eq. (6.106) leads to

$$D_A \frac{d^2 u}{dx^2} + c \frac{du}{dx} + ku(1-u) = 0. \tag{6.108}$$

This equation is non-linear, and we will not try to solve it. Rather, we will do a simple phase-plan analysis, and to this end we rewrite the equation as two coupled first-order differential equations:

$$\frac{du}{dx} = v \tag{6.109a}$$

$$D_A \frac{dv}{dx} = -cv - ku(1-u). \tag{6.109b}$$

We can identify two steady states, namely, $(u_1, v_1) = (0,0)$ and $(u_2, v_2) = (1,0)$, corresponding to pure B and A situations, respectively. To evaluate the stability, we linearise the system around each steady state and evaluate the eigenvalues of the corresponding Jacobi matrix:

$$\mathbf{J}(u,v) = \begin{bmatrix} 0 & 1 \\ k/D_A(u/2 - 1) & -c \end{bmatrix}. \tag{6.110}$$

For the point $(1,0)$ we have

$$\lambda_{1,2} = -\frac{1}{2} \left(\frac{c}{D_A} \pm \sqrt{\left(\frac{c}{D_A}\right)^2 + \frac{4k}{D_A}} \right). \tag{6.111}$$

Since $[c^2/D_A^2 + 4k/D_A]^{1/2} > c/D_A$, one eigenvalue will be positive and one negative, that is, the steady state $(1,0)$ is an unstable saddle point. The Jacobi matrix in $(0,0)$ has eigenvalues

$$\lambda_{1,2} = -\frac{1}{2} \left(\frac{c}{D_A} \pm \sqrt{\left(\frac{c}{D_A}\right)^2 - \frac{4k}{D_A}} \right). \tag{6.112}$$

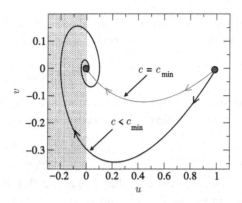

Two phase space trajectories (i.e. solutions) for Eq. (6.109); arrows indicate the solutions' flow direction. The shaded area is the forbidden region of negative concentration, $u = n_A/n_0$, and the filled circles the steady states.

If $(c/D_A)^2 < 4k/D_A$, the eigenvalues are complex (and conjugated), and the steady state is a stable spiral; see the phase plane in Fig. 6.18. However, this violates the physical constraint that $u \geq 0$. We must therefore require that the eigenvalues for the steady state $(0,0)$ has zero imaginary part, that is,

$$\left(\frac{c}{D_A}\right)^2 \geq \frac{4k}{D_A} \;\Rightarrow\; c \geq c_{\min} = 2\sqrt{kD_A}. \tag{6.113}$$

Thus, for $c \geq c_{\min}$ the origin is a stable node and $u \to 0$ for $z \to \infty$ and $u \to 1$ for $z \to -\infty$. As we have noted, from numerical integration of the Fisher–Kolmogorov equation, it has been shown that $c_{\min}$ is, in fact, the stable wave speed of the front.

In the very leading edge of the front, the concentration of A is small, and we can write $r \approx ku$ in this region. In the edge we then have

$$D_A \frac{d^2 u}{dx^2} + c\frac{du}{dx} + ku = 0, \tag{6.114}$$

which is a simple linear problem giving an exponential decay of u with respect to x; the corresponding eigenvalues are given by Eq. (6.112). In nanoscale chemical systems, the discrete nature of the molecules introduces a discontinuity in the reactive wave front, so the exponential decay is not obeyed. This is not an effect from fluctuations, but from the inherent discreteness of small-scale systems.

The discretisation introduces a cutoff, δ, in the reaction diffusion equation, defined by Brunet and Derrida [37]. A small number of reactive molecules, $n_{\rm rz}$, in the reaction zone corresponds to a large cutoff and vice versa; that is, we conjecture that $\delta = 1/n_{\rm rz}$ [95]. The particle number in the reaction zone is defined to be

$$n_{\rm rz} = nW_{\rm rz}A_{\rm rz}, \tag{6.115}$$

where n is the fluid number density, $W_{\rm rz}$ is a characteristic reaction zone width parallel to the front propagation direction, and $A_{\rm rz}$ the reaction zone area normal to the propagation; $A_{\rm rz}$ is in order of nanometres squared in our treatment here. Using the result from Brunet and Derrida and the relation between δ and n_{rz}, the effect of the inherent discrete nature of the system on the wave speed is summarised to

$$c_{rel} = \frac{c - c_{min}}{c_{min}} \propto \frac{1}{[\ln(\delta)]^2} = \frac{1}{[\ln(nW_{rz}A_{rz})]^2}. \tag{6.116}$$

In order to finalise our exploration, we need to specify the reaction zone width, W_{rz}. From a simple geometrical perspective, we define

$$W_{rz} = -\left.\frac{1}{du/dx}\right|_{x=0}, \tag{6.117}$$

where $x = 0$ is chosen such that $u(0) = 1/2$ [95]. With this definition we must have an expression for u, and again assuming the front exists and has a speed c, this is possible within some approximation using a perturbation method.

Defining the perturbation parameter, $\varepsilon = 1/c^2$, we can introduce the dimensionless spatial coordinate [162],

$$\zeta = k\sqrt{\varepsilon}x. \tag{6.118}$$

Application of the chain rule leads to

$$\varepsilon k D_A \frac{d^2u}{d\zeta^2} + \frac{du}{d\zeta} + u(1-u) = 0 \tag{6.119}$$

from Eq. (6.108). This form allows for a singular perturbation analysis, that is, we seek an approximate solution in the limit of small ε, or equivalently, in the limit of large front propagation speeds. We look for a solution of the form

$$u = u_0 + \varepsilon u_1 + \varepsilon^2 u_2 + \dots, \tag{6.120}$$

hoping that u is well approximated after a few terms. Inserting Eq. (6.120) into Eq. (6.119) and collecting the powers of ε, we have for the zeroth-order term

$$\frac{du_0}{d\zeta} + u_0(1 - u_0) = 0. \tag{6.121}$$

We specify the initial value shortly. Separation of variables gives

$$\frac{du_0}{u_0(u_0 - 1)} = d\zeta \quad \Rightarrow \quad u_0 = \frac{1}{1 - Ke^\zeta}, \tag{6.122}$$

where K is a constant of integration. The coordinate system origin can be chosen arbitrarily; here we choose the origin such that $u_0(0) = 1/2$, that is, the point where half of the B molecules are converted into A in accordance with the preceding discussion. We then have

$$u_0 = \frac{1}{1 + e^\zeta} = \frac{1}{1 + e^{kx/c}}. \tag{6.123}$$

Interestingly, the zeroth-order approximation is within 5 per cent of the numerical solution, and we will not consider higher-order terms from here on. The slope at $x = 0$ is found to be

$$\left.\frac{du}{dx}\right|_{x=0} = -\frac{k}{4c}, \tag{6.124}$$

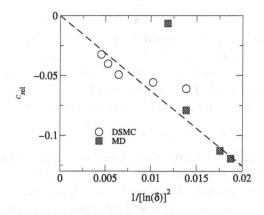

Figure 6.19 Relative wavefront speed reduction as a function of $1/[\ln(\delta)]^2$, using $1/\delta = 8nA_{rz}\sqrt{D_A/k}$. Symbols represent results from the master equation (open circles) and molecular dynamics simulations in the dilute case, $n = 0.2$ (filled squares). Punctured line has slope $-\pi^2/2$ [37]. Data are re-plotted from Ref. [95].

implying that

$$W_{rz} = \frac{4c}{k}. \qquad (6.125)$$

Inserting the minimum speed solution, $c = c_{\min}$, gives the expression for the reaction zone width in terms of the rate constant and the diffusion constant $W_{\min} = 8\sqrt{D_A/k}$. Substituting into Eq. (6.116), we get the final result for the effect of the molecular discreteness on the front speed:

$$c_{rel} \propto \frac{1}{\left[\ln(8nA_{rz}\sqrt{D_A/k})\right]^2}. \qquad (6.126)$$

Figure 6.19 plots the relative wavefront speed reduction, c_{rel}, as a function of the cut-off, $1/\delta = 8nA_{rz}\sqrt{D_A/k}$. The symbols represent simulation data, and the punctured line is the Brunet–Derrida prediction, $c_{rel} \propto 1/[\ln(\delta)]^2$. Note that the wave speed is significantly reduced for the small reaction zone volumes, $A_{rz}\sqrt{D_A/k}$; hence, this effect is important for nanoscale chemical wavefronts.

6.4 Further Explorations

1. Derive the one-dimensional heat equation, Eq. (6.2).

2. Recall that for the slit-pore geometry the kinetic temperature balance equation can be written as

$$\frac{dJ^\varepsilon}{dz} = -P_{xz}\frac{du_x}{dz}$$

in the steady state. Derive the temperature profile for a Couette flow in a slit-pore with two identical walls, $L_s^{(1)} = L_s^{(2)} = L_s$, using Fourier's constitutive relation $J^\varepsilon = -\lambda \, dT/dz$.

How does the temperature gradient depend on the slip length L_s? Compare with simulations of a Lennard–Jones liquid.

Discuss the extended constitutive relation by Baranyai et al. for the Couette flow.

Computational resources available.

3. In Section 6.2.4 the slip enhancement coefficient E^{slip} for electro-osmotic flows was plotted as a function of Debye length, λ_D. Derive the analytical expression for E^{slip}.
4. In this exploration we study the counter-ion system described in Section 6.2.2. First, derive the electric potential function of a counter-ion solution in the Debye–Hückel regime if the solution is confined between two charged surfaces, and

$$\varphi_q(0) = \varphi_q(h) = \zeta \, .$$

Then, derive the charge density profile. How does this differ from an electrolyte solution, Eq. (6.77)?

Finally, derive the equation for the corresponding electro-osmotic flow. Again, consider how this differs from an electrolyte solution, Eq. (6.84).

Epilogue

The continuum picture on which hydrodynamics is historically based stems from our macroscopic experience of smooth fluid flows. Hydrodynamics is therefore often regarded, rightly, as a macroscopic theory. Whether the same hydrodynamic equations are applicable on the nanoscale, where we approach the length scale of the fluid's intrinsic discrete nature, is not trivial. Specifically, we can pose the research question, 'Are the underlying physical processes dominant on the macroscopic scale also the dominant processes on the nanoscopic length scale?'. This question is paramount if we wish to model, that is, understand, predict, and control nanoscale fluid systems.

Onsager's regression hypothesis has been leading the way to connect the (average) relaxation of microscopic thermally induced perturbations to macroscopic hydrodynamics. This microscopic picture was used in Chapter 2 to derive the hydrodynamic balance equations. From these fundamental equations we revisited the established theory on thermal relaxation dynamics in Chapter 3, focusing on simple point-mass-type fluids modeled by the Lennard–Jones fluid. The main point of the chapter was that classical hydrodynamics can, in fact, account for the governing physical processes on the nanoscale for these systems. An important conclusion of the analysis was that the agreement depends on the actual dynamics and system we study. For example, the hydrodynamic prediction for the dispersion relation for the Brillouin peak fails at wavevectors, where the transverse dynamics follows the predictions (down to just a few nanometres).

As the system complexity increases, or as the length scale decreases below a few nanometres, the classical hydrodynamic theory will eventually break down. However, the hydrodynamic model can be extended, for example, by replacing Newton's law of viscosity with the Maxwell viscoelastic model or by introducing a wavevector-dependent viscosity. This was described in Chapter 4. Interestingly, the Maxwell model originates from macroscopic modelling of complex molecular systems; but it is the same viscoelastic effects that govern simple fluids at small time and length scales.

In Chapters 4–6 we explored different coupling phenomena, for example, the spin angular momentum and linear momentum coupling, the kinetic temperature and strain rate coupling, and the coupling between the flow profile and charge density distribution. In all cases the hydrodynamic theory can be extended, and from this we gain fundamental new knowledge of these coupling phenomena and how they affect the systems' dynamics on the nanoscale.

Chapters 3 and 5 showed the fluid structure on small scales. For example, fluid packing in equilibrium Lennard–Jones systems, Fig. 3.2, and molecular alignment of a butane liquid in a slit-pore, Fig. 5.4. Hydrodynamics, as it is defined here, cannot account for this structuring. However, while a strong structuring can affect the fluid dynamics, this effect is smaller than one may expect at first for the simple fluid systems we explored. Studying the structural effects for more complex fluids, like glasses, calls for different and further extended theoretical frameworks like the Mori–Zwanzig projection formalism and mode coupling theory.

The statement *"Hydrodynamics fails ..."* is not really meaningful. First, hydrodynamics is based on models, so it is by definition never exact, but always an approximation – even on the macroscopic scale. Secondly, if we seek to test the approximation, we must specify

  (i) what system we explore,
 (ii) the dynamical phenomenon,
(iii) how we quantify and define 'failure', and
(iv) exactly what we mean by hydrodynamics.

An absurd example; if we, strictly, do not allow for fluctuating hydrodynamic variables, then hydrodynamics is valid only in the thermodynamic limit of infinite system size.

Currently, research on nanoscale hydrodynamics relies heavily on testing the theory against atomistic computer simulations. This approach, is very appealing as it offers a great deal of control; however, it can only be as precise as our model of the molecular interactions, wall properties, and so on. Furthermore, the simulations are severely limited with respect to both time and length scales, reducing the phenomenology we can access. Nevertheless, as in all of science, computer simulations will play an ever-increasing role in our understanding of fluid dynamics. The underlying algorithms and models that are behind the results need careful attention, and one should always have access to the computer code in order to perform a fully informed research exploration and in case of result disagreements. Most importantly, we must always consider simulations as guidelines for the design of the experiments that ultimately corroborate the theories.

Appendix: A Bit of Help

A.1 The Fourier Transform

The reader is likely familiar with the Fourier series and the Fourier transform in one dimension. This appendix mainly seeks to clarify the symbolism which is used in the text, but also to briefly introduce the three dimensional Fourier transform which may not be familiar to the reader. For further information on the topic, there are very many good resources available on the internet. The careful and elegant treatment by Tolstov [204] is highly recommended.

In the book we explore, for example, the mass density. This is a scalar function of both time and space. We can study such a function over time at some point or at some specific time and as a function of a spatial coordinate; see the illustration in Fig. A.1. In either case, in general we can consider a periodic function (of one-dimensional space or time) f, where $f : \mathbb{R} \to \mathbb{R}$. The period of the function is denoted L (see illustration). Assume that f can be represented by an infinite trigonometric series

$$f(x) = \frac{a_0}{2} + \sum_{n=1}^{\infty} a_n \sin(k_n x) + b_n \cos(k_n x),$$

where a_n and b_n are the term amplitudes, known as Fourier coefficients, and $k_n = n\pi/L$ is the wavevector. This is the Fourier series.

The Fourier series can be written in an (elegant) complex form. From Euler's identity we have

$$\cos(k_n x) = \frac{e^{ik_n x} + e^{-ik_n x}}{2}$$

$$\sin(k_n x) = \frac{e^{ik_n x} - e^{-ik_n x}}{2i}.$$

Inserting this into the Fourier series, we get

$$f(x) = \frac{a_0}{2} + \sum_{n=1}^{\infty} \frac{a_n}{2}(e^{ik_n x} + e^{-ik_n x}) + \sum_{n=1}^{\infty} \frac{b_n}{2i}(e^{ik_n x} - e^{-ik_n x})$$

$$= \frac{a_0}{2} + \sum_{n=1}^{\infty} \frac{a_n - ib_n}{2}e^{ik_n x} + \sum_{n=1}^{\infty} \frac{a_n + ib_n}{2}e^{-ik_n x}.$$

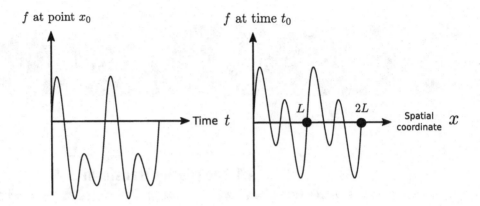

Illustration of a periodic function in either time or space. L is the function fundamental period with respect to spatial coordinate x.

Using that $-k_n = k_{-n}$, we can let the index run from $-\infty$ to ∞, and obtain

$$f(x) = \sum_{n=-\infty}^{\infty} c_n e^{ik_n x},$$

where c_n is the complex Fourier coefficient which is a function of a_n and b_n. We will not show this, but in general the complex Fourier coefficient is given by

$$c_n = \frac{1}{2L} \int_{-L}^{L} f(x) e^{-ik_n x} \, dx.$$

For non-periodic functions, we still define the Fourier coefficient by letting the period go infinite:

$$c_n = \lim_{L \to \infty} \frac{1}{2L} \int_{-L}^{L} f(x) e^{-ik_n x} \, dx. \tag{A.1}$$

This now leads to our definition of the Fourier transform: Let $f : \mathbb{R} \to \mathbb{R}$ be absolutely integrable, that is,

$$\int_{-\infty}^{\infty} |f(x)| \, dx$$

exists, then the Fourier transform of f is

$$\mathcal{F}[f] = \widetilde{f}(k) = \int_{-\infty}^{\infty} f(x) e^{-ikx} \, dx \tag{A.2}$$

for $k \in \mathbb{R}$. In the text we refer to $\widetilde{f}(k)$ as the Fourier coefficient or Fourier mode amplitude.

The Fourier transform is not uniquely defined! From our definition, the inverse Fourier transform is given by

$$\mathcal{F}^{-1}[\widetilde{f}] = \frac{1}{2\pi} \int_{-\infty}^{\infty} \widetilde{f}(k) e^{ikx} \, dk$$

such that $\mathcal{F}^{-1}[\mathcal{F}[f]] = \mathcal{F}^{-1}[\widetilde{f}] = f(x)$.

These ideas can be extended to three dimensions. Here we have $f : \mathbb{R}^3 \to \mathbb{R}$, and if f is absolutely integrable, that is,

$$\iiint_{-\infty}^{\infty} |f(x,y,z)| \, dxdydz$$

exists, we define the Fourier transform of f as

$$\mathcal{F}[f] = \tilde{f}(\mathbf{k}) = \iiint_{-\infty}^{\infty} f(x,y,z) e^{-i(k_x x + k_y y + k_z z)} \, dxdydz,$$

where $\mathbf{k} = (k_x, k_y, k_z) \in \mathbb{R}^3$. In the text we write this in compact notation. Let $\mathbf{r} = (x,y,z)$ then,

$$\mathcal{F}[f] = \int_{-\infty}^{\infty} f(\mathbf{r}) e^{-i\mathbf{k}\cdot\mathbf{r}} \, d\mathbf{r}.$$

In three dimensions, the inverse Fourier transform is (in compact notation)

$$\mathcal{F}^{-1}[\tilde{f}] = \frac{1}{(2\pi)^3} \int_{-\infty}^{\infty} \tilde{f}(\mathbf{k}) e^{i\mathbf{k}\cdot\mathbf{r}} \, d\mathbf{k}$$

such that, again, $\mathcal{F}^{-1}[\mathcal{F}[f]] = \mathcal{F}^{-1}[\tilde{f}] = f(\mathbf{r})$.

Example Let f be the three-dimensional Gaussian function

$$f(x,y,z) = e^{-(x^2+y^2+z^2)}.$$

Notice that the domain of f is all $\mathbb{R}^3$ and that f is absolutely integrable. The three-dimensional Fourier transform is then, from the definition,

$$\mathcal{F}[f] = \tilde{f}(\mathbf{k}) = \int_{-\infty}^{\infty} e^{-(x^2+y^2+z^2)} e^{-i(k_x^2+k_y^2+k_z^2)} \, d\mathbf{r}.$$

The integrant is separable, and we can write it as the product of three functions,

$$\tilde{f}(\mathbf{k}) = \int_{-\infty}^{\infty} e^{-(x^2+ik_x x)} e^{-(y^2+ik_y y)} e^{-(z^2+ik_z z)} \, d\mathbf{r}$$

$$= (\sqrt{\pi})^3 e^{-k_x^2/4} e^{-k_y^2/4} e^{-k_z^2/4} = \pi^{3/2} e^{-k^2/4},$$

where $k^2 = \mathbf{k}\cdot\mathbf{k}$.

The inverse Fourier transformation of $\tilde{f}$ is (again using separation of the function)

$$\mathcal{F}^{-1}[\tilde{f}] = \frac{\pi^{3/2}}{(2\pi)^3} \int_{-\infty}^{\infty} e^{-(k_x^2+k_y^2+k_z^2)/4} e^{i(k_x x + k_y y k_z z)} \, d\mathbf{k}$$

$$= \frac{\pi^{3/2}}{(2\pi)^3} (2\sqrt{\pi})^3 e^{-x^2} e^{-y^2} e^{-z^2}$$

$$= e^{-(x^2+y^2+z^2)} = f(x,y,z).$$

Thus, the original function is recaptured.

A.2 The Dirac Delta

Like the Fourier transform, the reader is probably familiar with the Dirac delta in one dimension, and again, this appendix acts to clarify the symbolism and extend the idea of the Dirac delta from one dimension to three dimensions.

In one dimension, the Dirac delta fulfils the following two fundamental properties:

1. Let $x \in \mathbb{R}$

$$\delta(x) = \begin{cases} \infty & \text{if } x = 0 \\ 0 & \text{if } x \neq 0 \end{cases}$$

2. and

$$\int_{-\infty}^{\infty} \delta(x) \mathrm{d}x = 1.$$

Notice that the unit of the Dirac delta is the inverse of the argument's unit. We can now show a very important property of the Dirac delta, which we use frequently in the book. Let $f : \mathbb{R} \to \mathbb{R}$ be continuous; then, from property 1

$$\int_{-\infty}^{\infty} f(x) \delta(x-a) \mathrm{d}x = \int_{a-\varepsilon}^{a+\varepsilon} f(x) \delta(x-a) \mathrm{d}x \quad (\varepsilon > 0).$$

As $\varepsilon \to 0$, we have that f approaches some constant value, $f(a)$, and therefore

$$\lim_{\varepsilon \to 0} \int_{a-\varepsilon}^{a+\varepsilon} f(x) \delta(x-a) \mathrm{d}x = f(a) \lim_{\varepsilon \to 0} \int_{a-\varepsilon}^{a+\varepsilon} \delta(x-a) \mathrm{d}x = f(a).$$

That is,

$$\int_{-\infty}^{\infty} f(x) \delta(x-a) \mathrm{d}x = f(a).$$

The extension to three dimensions is straightforward, albeit the symbolism used in the book can appear somewhat subtle. The two fundamental properties are

1. Let $\mathbf{r} \in \mathbb{R}^3$, then

$$\delta(\mathbf{r}) = \delta(x)\delta(y)\delta(z) = \begin{cases} \infty & \text{if } x = 0 \text{ and } y = 0 \text{ and } z = 0 \\ 0 & \text{otherwise} \end{cases}$$

2. and

$$\iiint_{-\infty}^{\infty} \delta(x)\delta(y)\delta(z) \, \mathrm{d}x\mathrm{d}y\mathrm{d}z = \int_{-\infty}^{\infty} \delta(\mathbf{r})\mathrm{d}\mathbf{r} = 1,$$

using the compact notation introduced in A.1.

Following the idea from the one-dimensional case, we have for the continuous function $f : \mathbb{R}^3 \to \mathbb{R}$

$$\int_{-\infty}^{\infty} f(\mathbf{r})\delta(\mathbf{r}-\mathbf{a})d\mathbf{r} = \lim_{\varepsilon\to 0} \int_{\mathbf{a}-\varepsilon}^{\mathbf{a}+\varepsilon} f(\mathbf{r})\delta(\mathbf{r}-\mathbf{a})d\mathbf{r}$$

$$= f(\mathbf{a})\lim_{\varepsilon\to 0}\int_{\mathbf{a}-\varepsilon}^{\mathbf{a}+\varepsilon}\delta(\mathbf{r}-\mathbf{a})d\mathbf{r} = f(\mathbf{a}).$$

This result is used throughout the text for the exponential function, $f(\mathbf{r}) = e^{i\mathbf{k}\cdot\mathbf{r}}$, giving, for $\mathbf{a} = \mathbf{r}_i$,

$$\int_{-\infty}^{\infty} e^{i\mathbf{k}\cdot\mathbf{r}}\delta(\mathbf{r}-\mathbf{r}_i)d\mathbf{r} = e^{i\mathbf{k}\cdot\mathbf{r}_i}.$$

Example We finish with the standard example showing that the microscopic definition of a hydrodynamic variable in terms of the Dirac delta is not obtuse, even if the distribution features infinite peaks at the molecules' centre of mass. We will take the mass density

$$\rho(\mathbf{r},t) = \sum_i m_i\delta(\mathbf{r}-\mathbf{r}_i).$$

If we integrate the left-hand side over the system volume,

$$\iiint_{\mathcal{V}}\rho(\mathbf{r},t)dxdydz = \int_{\mathcal{V}}\rho(\mathbf{r},t)d\mathbf{r} = M_{\mathcal{V}},$$

we get the total mass in the system. Let us do the same on the right-hand side. Not being particularly concerned about convergence properties, we move the integration under the summation sign,

$$\int_{\mathcal{V}}\sum_i m_i\delta(\mathbf{r}-\mathbf{r}_i)d\mathbf{r} = \sum_i m_i\int_{\mathcal{V}}\delta(\mathbf{r}-\mathbf{r}_i)d\mathbf{r} = \sum_i m_i = M_{\mathcal{V}},$$

and we recapture the system mass, as expected.

A.3 Expected Value, Average, and Correlation Functions

In our explorations, the hydrodynamic quantities $A = \rho\phi$ are random functions, also referred to as random variables or random processes. The domain of these quantities is spanned by time and space. Let us simplify the discussion a bit by letting a random function, denoted x, be a scalar function. The expected value for x is defined by

$$\langle x\rangle = \int_{-\infty}^{\infty} xf(x)dx,$$

where f is the probability density function. Sometimes $E[x]$ is used as symbol for the expected value. The unit of f is the inverse of the unit of x. The expected value is also related to the so-called first moment of f; a measure of the function's "mass" midpoint.

Now, in practice, when we perform simulations, we sample the hydrodynamic quantities from a set of systems initialised at different initial conditions. If we have N such systems, we obtain an ensemble of N discrete independent measurements, $x_1, x_2, \ldots, x_N$. The expected value for the discrete ensemble is

$$\langle x\rangle = \sum_i x_i p_i.$$

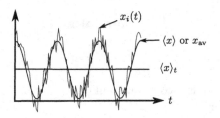

Figure A.2 Illustration of the different averages $\langle x \rangle$, x_{av}, and $\langle x \rangle_t$. x_i represents one sample in the sample ensemble, and $\langle x \rangle_t$ is the long time average $\tau \to \infty$.

Here p_i is the probability that x_i occurs. The probabilities obey $\sum_i p_i = 1$; thus, if we assume that all x_i occur with same probability, p, then $\sum_i p_i = Np = 1$, implying $p = 1/N$. The expected value is then simply the ensemble sample average

$$\langle x \rangle = x_{av} = \frac{1}{N} \sum_i x_i .$$

Importantly, we can form the ensemble average at different times, such that $\langle x \rangle(t) = \frac{1}{N} \sum_i x_i(t)$. We see that the ensemble sample average is not, in general, the same as the time average denoted $\langle x \rangle_t$, defined by

$$\langle x \rangle_t = \frac{1}{\tau} \int_0^\tau x(t) \mathrm{d}t .$$

Figure A.2 illustrates the differences between a member of the ensemble, $x_i(t)$, the expected value $\langle x \rangle$, the sample average x_{av}, and the long time average $\langle x \rangle_t$. Finally, we also encounter spatial averages in the text; this is indicated by an overline, $\bar{x}$. Thus if x is a function of y, we have

$$\bar{x} = \frac{1}{b-a} \int_a^b x \mathrm{d}y ,$$

where the limits a and b are defined by the actual problem.

A special case is equilibrium. From the ergodicity theorem, we have that for $N \to \infty$ and $\tau \to \infty$ [90]

$$\langle x \rangle = \langle x \rangle_t \text{ in equilibrium.}$$

That is, we can replace the expected value, or ensemble sample average, with the time average and vice versa in equilibrium.

Moving on to correlation functions. Let $x, y : I \to \mathbb{R}$, where $I = [0; \infty[$ represents the time interval. We here define the correlation function to be

$$C_{xy}(t) = \frac{1}{\tau} \int_0^\infty x(t'+t) y(t') \, \mathrm{d}t' , \tag{A.3}$$

where τ is a characteristic time that we must define and $t \geq 0$. We will not consider the necessary conditions for existence and boldly proceed. If $x \neq y$, the correlation function is referred to as the cross-correlation function, and if $x = y$, the correlation function C_{xx} is the autocorrelation function. The correlation function measures, hardly surprising, the correlation between x and y at different t-lags. That is, how x at time $t' + t$ is correlated with y at previous time t', on average over I.

The correlation function fulfils two important properties that we use in the book:

1. The cross-correlation function does not, in general, commute: $C_{xy} \neq C_{yx}$.
2. The autocorrelation function C_{xx} is even (when the domain is extended over] $-\infty; \infty[$).

In some definitions of the correlation function the t-lag pertains to the second function factor, and since the correlation function does not commute, one needs to be careful with the exact definition. Our definition follows the book theory, and we will stick with this. The two properties given here are independent of the exact definition.

We now arrive at a main point. From the definition we see that the correlation function is a time average of a product. For time translation invariance we write the correlation function $C_{xy}(t) = \langle x(t)y(0) \rangle_t$. Due to the ergodicity in equilibrium, the time average equals the ensemble average; that is,

$$C_{xy}(t) = \langle x(t)y(0) \rangle.$$

Therefore, when in equilibrium, we can calculate the correlation functions from simple ensemble averages of time series.

Example This example shows how to calculate the correlation function, but it also acts as a proof, by contradiction, of property 1. Let us propose (or assume) that the correlation function commutes, $C_{xy} = C_{yx}$. Define $x, y : I \to \mathbb{R}$ by

$$x(t) = e^{-t} \text{ and } y(t) = 1 + t.$$

Then, for $t \geq 0$ we have

$$C_{xy}(t) = \frac{1}{\tau} \int_0^\infty x(t' + t)y(t') \, dt'$$
$$= \frac{1}{\tau} \int_0^\infty e^{-(t'+t)}(1 + t') \, dt' = \frac{2e^{-t}}{\tau},$$

but

$$C_{yx}(t) = \frac{1}{\tau} \int_0^\infty y(t' + t)x(t') \, dt'$$
$$= \frac{1}{\tau} \int_0^\infty (1 + t' + t)e^{-t'} \, dt'$$
$$= \frac{1+t}{\tau} \int_0^\infty e^{-t'} \, dt' + \frac{1}{\tau} \int_0^\infty t'e^{-t'} \, dt' = \frac{2+t}{\tau},$$

that is, $C_{xy} \neq C_{yx}$, which contradicts the proposition.

A.4 Method of Undetermined Coefficients

Our explorations usually end with the problem of solving linear differential equations. Many of these equations are integrable or homogeneous, and the reader is likely familiar with solving these equations already. However, a few times we also encounter inhomogeneous differential equations which are, perhaps, not always straightforward to deal with. There are different ways to approach these problems, but one simple technique is the method of undetermined coefficients; and while this method will not always be applicable, it works for the problems we encounter. If the reader needs additional resources, the book by Boyce and DiPrima [34] is strongly recommended.

First-Order Differential Equations

Consider the initial value problem

$$\frac{dx}{dt} + ax = b \text{ with } x(0) = 0,$$

where $a, b \neq 0$ are constants. This is a linear first-order inhomogeneous differential equation; b is called the inhomogeneous part. Notice, here one cannot simply perform separation of variables and integrate; well, not unless we first find a helpful variable transformation.

The method of undetermined coefficients is based on the superposition principle. For this problem, it states that the solution, x, is a sum of the homogeneous solution, x_h, and the particular solution, x_p; that is,

$$x = x_h + x_p.$$

Thus, we need to find x_h and x_p.

Homogeneous solution The corresponding homogeneous differential equation is when $b = 0$, that is,

$$\frac{dx_h}{dt} + ax_h = 0,$$

which has the solution $x_h(t) = K_1 e^{-at}$.

Particular solution To find the particular solution, we try to guess a solution, and see if this can fulfil the differential equation. If the right-hand side of the differential equation is a polynomial, we guess that x_p is a polynomial of the same degree as the differential equation, here first order. Therefore, our guess is

$$x_p = At + B.$$

Substituting into the original differential equation and rearranging a bit, we obtain

$$(aA)t + (A + aB - b) = 0,$$

which must be true for all t we study (defined from the original differential equation). The equation is then fulfilled if each term is zero. Since $a \neq 0$ implies $A = 0$, therefore $B = b/a$, leading to $x_p = b/a$.

The general solution is $x = K_1 e^{-at} + b/a$. By applying the initial conditions, K_1 can be determined, resulting in

$$x(t) = \frac{b}{a}\left(1 - e^{-at}\right).$$

Note, had the inhomogeneous part been, say, a trigonometric function, we would guess that x_p was given by a sine, cosine, a linear combination of the two, and so forth; see Ref. [34].

More elegantly: it is possible to find an integrating factor for the general case where $a = a(t)$ and $b = b(t)$. This gives an integral equation solution to any linear first-order differential equation. Thus, we need not guess a particular solution x_p using this approach.

Second-Order Differential Equations

The method of undetermined coefficients for linear second-order differential equations follows the exact same idea as for first-order problems. However, the reader may find it useful to see an example of a second-order differential equation, as this will also treat the homogeneous problem, which we deal with very frequently.

We will simply go through a specific example which is of relevance for the type of problems we encounter in the text; let

$$\frac{d^2 x}{dt^2} + \frac{dx}{dt} - 2x = 2t.$$

We seek the general solution. According to the preceding discussion, we need a homogeneous and a particular solution.

Homogeneous solution The corresponding homogeneous equation is

$$\frac{d^2 x_h}{dt^2} + \frac{dx_h}{dt} - 2x_h = 0.$$

This has the characteristic function

$$\lambda^2 + \lambda - 2 = 0 \implies \lambda = \{1, -2\}$$

and therefore the homogeneous solution is

$$x_h(t) = K_1 e^t + K_2 e^{-2t}.$$

Particular solution The inhomogeneous part is a polynomial, and the differential equation is of second order. We therefore guess that the particular solution is a polynomial of second order:

$$x_p = At^2 + Bt + C.$$

If we substitute this into the differential equation, we get

$$-2At^2 + 2(A - B - 1)t + (B + 2A - 2C) = 0.$$

Equating each term with zero, we see that $A = 0$, $B = -1$, and $C = -1/2$, hence, $x_p = -2t + 1$. The general solution reads

$$x(t) = K_1 e^t + K_2 e^{-2t} - t - 1/2,$$

where the constants K_1 and K_2 are found from the problem boundary values or initial values.

A.5 Computational Resources

The book's additional resources include a molecular dynamics simulation package. The package consists of a molecular dynamics kernel library,[1] as well as a GNU Octave/Matlab wrapper. With the wrapper, the front-end user can easily access the functionality from the kernel library, as well as the built-in auxiliary functionality that GNU Octave and Matlab offer. On the book's web page a few introductory videos on the installation and use of the software package can be found. It is recommended that the reader consult these before using the software. Importantly, the software can run on a simple desktop or a laptop (depending on one's patience) and is, under the Linux operating system, relatively easy to install and to get started with.

The *Further Explorations* section at the end of each chapter contains a series of standard well-defined exercises and more open problems that the reader can engage. The open problems sometimes involve analysis of molecular dynamics data, and these data must be produced by the readers themselves. It is recommended to use the molecular dynamics software package that comes with the book. Table A.2 lists a set of GNU Octave/Matlab scripts and what *Further Explorations* they are designed for. Sometimes the script can be run directly without modifications; other times, the reader must make a few changes or even code some new functionality in order to explore the problem. The scripts are provided with a few comments that can help set up and run the simulations, as well as analyse the results.

The output of the simulations are given in molecular dynamics units. In the book we have sometimes presented results in these units some times in SI units. To quickly convert the simulation results to SI units (and vice versa), Table A.2 lists a series of relevant conversion factors for the methane and butane systems; this can be very convenient when analysing the output from the simulations. Conversion factors for any other molecular model can be derived directly using the table's second column and from the model Lennard–Jones parameter σ, ε, and m.

---

[1] Developed by the author.

Table A.1 List of scripts and what *Further Explorations* they are designed for. The auxiliary files are function files that are used in the data analysis, and are also part of the software package.

Script	Exploration	Input files	Aux. files
run_LJ.m	3.4	start_LJ.xyz	evcorr.mex or evcorr.oct
run_KA.m	3.5	start_KA.xyz	fltrans.m hann.m trapz.m
run_butane.m	4.1	start_butane.xyz start_butane.top	
run_toluene.m	4.2	start_toluene.xyz start_toluene.top	fltrans.m hann.m trapz.m
run_diatomic.m	4.3 & 4.4	start_diatomic.xyz start_diatomic.top	fltrans.m hann.m trapz.m
run_Cflow.m	5.1, 5.2 & 6.2	start_Cflow.xyz	

Table A.2 Lennard–Jones unit and conversion factors for methane and butane liquids.

Property	Lennard–Jones unit	Methane-factor	Butane-factor
Length	σ	3.70×10^{-10} m	3.90×10^{-10} m
Mass	m	2.67×10^{-26} kg	2.41×10^{-26} kg
Energy	ε	2.04×10^{-21} J	9.94×10^{-21} J
Temperature	ε/k_B	148 K	72 K
Time	$\sigma\sqrt{m\varepsilon^{-1}}$	1.33×10^{-12} s	1.92×10^{-12} s
Mass density	$m\sigma^{-3}$	524.52 kg m^{-3}	405.91 kg m^{-3}
Force	ε/σ	5.52×10^{-12} N	2.55×10^{-12} N
Pressure	ε/σ^{-3}	40.34 MPa	16.76 MPa
Viscosity	$\sqrt{m\varepsilon}/\sigma^2$	5.38×10^{-5} Pa· s	3.22×10^{-5} Pa· s
Diffusion	$\sigma/\sqrt{m\varepsilon^{-1}}$	1.03×10^{-7} m^2 s^{-1}	7.92×10^{-8} m^2 s^{-1}

References

[1] P. Abgrall. *Nanofluidics*. Artech House, Norwood, 2009.

[2] P. Lunkenheimer, S. Emmert, R. Gulich et al. Electromagnetic-radiation absorption of water. 96(6):062607, 2017.

[3] N. K. Ailawadi, B. J. Berne, and D. Forster. Hydrodynamics and collective angular-momentum fluctuations in molecular fluids. *Phys. Rev. A*, 3(4):1462–1472, 1971.

[4] M. P. Allen. Atomic and molecular representations of molecular hydrodynamic variables. *Mol. Phys.*, 52(3):705–716, 1984.

[5] M. P. Allen and D. J. Tildesley. *Computer Simulation of Liquids*. Clarendon Press, New York, 1989.

[6] W. E. Alley and B. J. Alder. Generalized transport coefficients for hard spheres. *Phys. Rev. A*, 27(6):3158, 1983.

[7] S. Alosious, S. K. Kannam, S. P. Sathian, and B. D. Todd. Prediction of Kapitza resistance at fluid-solid interfaces. *J. Chem. Phys.*, 151(19):194502, 2019.

[8] J. Badur, P. Ziolkowski, and P. Ziolkowski. On the angular velocity slip in nano-flows. *Microfluid. Nanofluidics*, 19:191–198, 2015.

[9] N. P. Bailey, T. S. Ingebrigtsen, J. S. Hansen et al. RUMD: A general purpose molecular dynamics package optimized to utilize GPU hardware down to a few thousand particle. *SciPost Phys.*, 3:038, 2017.

[10] V. Ballenegger and J. P. Hansen. Dielectric permittivity profiles of confined polar fluids. *J. Chem. Phys.*, 122(11):114711, 2005.

[11] U. Balucani and M. Zoppi. *Dynamics of the Liquid State*. Cambridge University Press, New York, 1994.

[12] A. Baranyai, D. J. Evans, and P. J. Daivis. Isothermal shear-induced heat-flow. *Phys. Rev. A*, 46(12):7593, 1992.

[13] F. Baras and M. M. Mansour. Microscopic simulations of chemical instabilities. *Adv. Chem. Phys.*, 100:393, 1997.

[14] J.-L. Barrat and F. Chiaruttini. Kapitza resistance at the liquid solid interface. *Mol. Phys.*, 101:1605, 2003.

[15] G. K. Batchelor. *An Introduction to Fluid Dynamics*. Cambridge University Press, 1967.

[16] S. H. Behrens and D. G. Grier. The charge of glass and silica surfaces. *J. Chem. Phys.*, 115(14):6716, 2001.

[17] D. Ben-Yaakov, D. Andelman, D. Harries, and R. Podgornik. Beyond standard Poisson–Boltzmann theory: ion-specific interactions in aqueous solutions. *J. Phys.: Condens. Matter*, 21(42):424106, 2009.

[18] S. Bernardi, B. D. Todd, and D. J. Searles. Thermostating highly confined fluids. *J. Chem. Phys.*, 132(24):244706, 2010.

[19] D. Bertolini and A. Tani. Generalized hydrodynamics and the acoustic modes of water: theory and simulation result. *Phys. Rev. E*, 51(2):1091, 1995.

[20] G. A. Bird. *Molecular Gas Dynamics and Direct Simulation of Gas Flows*. Oxford University Press, 1994.

[21] I. Bitsanis, J. J. Magda, M. Tirrell, and H. T. Davis. Molecular dynamics of flow in micropores. *J. Chem. Phys.*, 87(3):1733, 1987.

[22] I. Bitsanis, T. K. Vanderlick, M. Tirrell, and H. T. Davis. Tractable molecular theory of flow in strongly inhomogeneous fluids. *J. Chem. Phys.*, 89(5):3152, 1988.

[23] L. Bocquet and J.-L. Barrat. Hydrodynamic boundary conditions, correlation functions, and Kubo relations for confined fluids. *Phys. Rev. E*, 49(4):3079, 1994.

[24] L. Bocquet and J. L. Barrat. On the Green-Kubo relationship for the liquid-solid friction coefficient. *J. Chem. Phys.*, 139(4):044704, 2013.

[25] L. Bocquet and E. Charlaix. Nanofluidics, from bulk to interface. *Chem. Soc. Rev.*, 39(3):1073, 2010.

[26] L. Bocquet and P. Tabeling. Physics and technological aspects of nanofluidics. *Lab on a Chip*, 14(17):3143, 2014.

[27] T. Bodensteiner, C. Morkel, W. Gläser, and B. Dorner. Collective dynamics in liquid cesium near the melting point. *Phys. Rev. A*, 45(8):5709, 1992.

[28] J. D. Bonthuis, D. Horinek, L. Bocquet, and R. R. Netz. Electrohydraulic power conversion in planar nanochannels. *Phys. Rev. Lett.*, 103(14):144503, 2009.

[29] J. D. Bonthuis and R. R. Netz. Unraveling the combined effects of dielectric and viscosity profiles on surface capacitance, electro-osmotic mobility, and electric surface conductivity. *Langmuir*, 28(46):16049, 2012.

[30] J. D. Bonthuis, S. Gelke, and R. R. Netz. Profile of the static permittivity tensor of water at interfaces: consequences for capacitance, hydration interaction and ion adsorption. *Langmuir*, 28(20):7679, 2012.

[31] J. P. Boon and S. Yip. *Molecular Hydrodynamics*. Dover Publications, Mineola, NY, 1991.

[32] P. A. Bopp, A. A. Kornyshev, and G. Sutmann. Frequency and wave-vector dependent dielectric function of water: collective modes and relaxation spectra. *J. Chem. Phys.*, 109(5):1939, 1998.

[33] I. C. Bourg and G. Sposito. Molecular dynamics simulations of the electrical double layer on smectite surfaces contacting concentrated mixed electrolyte ($NaCl$–$CaCl_2$) solutions. *J. Coll. Inter. Science*, 360(2):701, 2011.

[34] W. E. Boyce and R. C. DiPrima. *Elementary Differential Equations and Boundary Value Problems*. John Wiley & Sons, New York, 1997.

[35] F. Bresme, A. Lervik, D. Bedaux, and S. Kjelstrup. Water olarization under thermal conditions. *Phys. Rev. Lett.*, 101(2):020602, 2008.

[36] B. R. Brooks, R. E. Bruccoleri, B. D. Olafson et al. CHARMM: a program for macromolecular energy, minimization, and dynamics calculations. *J. Comp. Chem.*, 4(2):187, 1983.

[37] E. Brunet and B. Derrida. Shift in the velocity of a front due to a cutoff. *Phys. Rev. E*, 56(3):2597, 1997.

[38] H. Bruus. *Theoretical Microfluidics*. Oxford University Press, Oxford, 2008.

[39] T. Bryk, I. Mryglod, G. Ruocco, and T. Scopigno. Comment on 'Emergence and evolution of the k gap in spectra of liquid and supercritical states'. *Phys. Rev. Lett.*, 120(21):219601, 2018.

[40] T. Bryk, I. Mryglod, T. Scopigno et al. Collective excitations in supercritical fluids: analytical and molecular dynamics study of 'positive' and 'negative' dispersion. *J. Chem. Phys.*, 133:024502, 2010.

[41] P. J. Cadusch, B. D. Todd, J. Zhang, and P. J. Daivis. A non-local hydrodynamic model for the shear viscosity of confined fluids: analysis of homogeneous kernel. *J. Phys. A*, 41(3):035501, 2008.

[42] S. Chen and G. D. Doolen. Lattice Boltzmann method for fluid flows. *Annu. Rev. Fluid Mech.*, 30:329, 1998.

[43] E. Cosserat and F. Cosserat. Sur la théorie de l'esticité. *Ann. Toulouse*, 10:1, 1896.

[44] E. Cosserat and F. Cosserat. *Théorie des corps d'eformables*. Hermann, 1909.

[45] F. Croccolo, J. M. Ortiz de Zárate, and J. V. Sengers. Non-local fluctuation phenomena in liquids. *Eur. Phys. Journ. E*, 39:125, 2016.

[46] J. S. Dahler and L. E. Scriven. Theory of structured continua I. General considerations of angular momentum and polarization. *Proc. R. Soc. Lond. A*, 27(1363):504, 1963.

[47] P. J. Daivis and D. J. Evans. Comparison of constant pressure and constant volume nonequilibrium simulations of sheared model decane. *J. Chem. Phys.*, 100(1):541–547, 1994.

[48] P. J. Daivis and J. L. Khayyam Coelho. Generalized Fourier law for heat flow in a fluid with a strong, nonuniform strain rate. *Phys. Rev. E*, 61(58):6003, 2000.

[49] B. A. Dalton, P. J. Daivis, J. S. Hansen, and B. D. Todd. Effects of nanoscale density inhomogeneities on shearing fluids. *Phys. Rev. E*, 88(5):052143, 2013.

[50] O. Darrigol. *Worlds of Flow: A History of Hydrodynamics from the Bernoullis to Prandtl*. Cambridge University Press, Cambridge, UK, 2005.

[51] P. G. de Gennes. Liquid dynamics and inelastic scattering of neutrons. *Physica*, 25(7–12):825, 1959.

[52] S. R. de Groot and P. Mazur. *Non-equilibrium Thermodynamics*. Dover Publications, Mineola, NY, 1984.

[53] S. R. de Groot and L. G. Suttorp. *Foundations of Electrodynamics*. North-Holland Publishing, Amsterdam, 1972.

[54] I. M. de Schepper, E. G. D. Cohen, C. Bruin et al. Hydrodynamic time correlation functions for a Lennard-Jones fluid. *Phys. Rev. A*, 38(1):271, 1988.

[55] J. M. Ortiz de Zárate, R. Pérez Cordón, and J. V. Senger. Finite-size effects on fluctuations in a fluid out of thermal equilibrium. *Physica A*, 291(1):113, 2001.

[56] J. M. O. de Zárate and J. V. Sengers. *Hydrodynamic Fluctuations*. Elsevier, Amsterdam, 2006.

[57] D. I. Dimitrov, A. Milchev, and K. Binder. Capillary rise in nanopores: molecular dynamics evidence for the Lucas-Washburn equation. *Phys. Rev. Lett.*, 99(5):054501, 2007.

[58] O. V. Dolgov, D. A. Kirzhnits, and E. G. Maksimov. On an admissible sign of the static dielectric function of matter. *Rev. Mod. Phys.*, 53(1):81, 1981.

[59] M. F. Döpke, J. Lützenkirchen, O. A. Moultos et al. Preferential adsorption in mixed electrolytes confined by charged amorphous silica. *J. Phys. Chem.*, 123(27):16711, 2019.

[60] D. M. F. Edwards, P. A. Madden, and I. R. McDonald. A computer simulation study of the dielectric properties of a model of methyl cyanide. *Mol. Phys.*, 51(5):1141, 1984.

[61] J. C. T. Eijkel and A. van den Berg. Nanofluidics: what is it and what can we expect from it? *Microfluidics Nanofluidics*, 1(3):249–267, 2005.

[62] A. C. Eringen. Mechanics of micropolar continua. In *Contributions to Mechanics. Markus Reiner Eightieth Anniversary Volume*, pp. 23–40. Edited by D. Abir. Pergamon, Oxford, 1969.

[63] P. Español and P. Warren. Statistical mechanics of dissipative particle dynamics. *Europhys. Lett.*, 30(4):191, 1995.

[64] D. J. Evans and H. J. M. Hanley. Fluctuation expressions for fast thermal processes: vortex viscosity. *Phys. Rev. A.*, 25(3):1771–1774, 1982.

[65] D. J. Evans and G. P. Morris. Nonlinear-response theory for steady planar Couette flow. *Phys. Rev. A.*, 30(3):1528, 1984.

[66] D. J. Evans and G. P. Morriss. *Statistical Mechanics of Nonequilibrium Liquids*. Academic Press, Cambridge, MA, 2008.

[67] D. J. Evans and W. B. Streett. Transport properties of homonuclear diatomics II. Dense fluids. *Mol. Phys.*, 36(1):161–176, 1978.

[68] J. P. Ewen, D. M. Heyes, and D. Dini. Advances in nonequilibrium molecular dynamics simulations of lubricants and additives. *Friction*, 6:349, 2018.

[69] J. Fan and L. Wang. Review of heat conduction in nanofluids. *J. Heat Trans.*, 133(4):040801, 2011.

[70] M. Fedoruk, M. Meixner, S. Carretero-Palacios, T. Lohmüller, and J. Feldmann. Nanolithography by plasmonic heating and optical manipulation of gold nanoparticles. *ACS Nano*, 7(9):7648, 2013.

[71] B. U. Felderhof. Efficiency of magnetic plane wave pumping of a ferrofluid through a planar duct. *Phys. Fluids*, 23(9):092003, 2011.

[72] B. U. Felderhof. Self-propulsion of a planar electric or magnetic microbot immersed in a polar viscous fluid. *Phys. Rev. E*, 83(5):056315, 2011.

[73] R. A. Fisher. The wave of advance of advantageous genes. *Ann. Eugenics*, 7(4):335, 1937.

[74] D. Frenkel and B. Smit. *Understanding Molecular Simulation*. Academic Press, London, 1996.

[75] J. Frenkel. *Kinetic Theory of Liquids*. Dover Publications, Mineola, 1955.

[76] L. Fumagalli, A. Esfandiar, R. Fabregas et al. Anomalously low dielectric constant of confined water. *Science*, 360(6395):1339, 2018.

[77] A. Furukawa and H. Tanaka. Nonlocal nature of the viscous transport in super-cooled liquids: complex fluid approach to supercooled liquids. *Phys. Rev. Lett.*, 103(13):135703, 2009.

[78] P. G. D. Gennes and J. Prost. *The Physics of Liquid Crystals*. Clarendon Press, Oxford, 1993.

[79] N. Gnan, T. B. Schrøder, U. R. Pedersen, N. P. Bailey, and J. C Dyre. Pressure-energy correlations in liquids. iv. 'Isomorphs' in liquid phase diagrams. *J. Chem. Phys.*, 131(23):234504, 2009.

[80] E. M. Gosling, I. R. McDonald, and K. Singer. On the calculation by molecular dynamics of the shear viscosity of a simple fluid. *Mol. Phys.*, 26(6):1475, 1973.

[81] W. Götze. *Complex Dynamics of Glass-Forming Liquids: A Mode-Coupling Theory*. Oxford Science Publications, Oxford, 2009.

[82] A. O. Govorov, W. Zhang, T. Skeini et al. Gold nanoparticle ensembles as heaters and actuators: melting and collective plasmon resonances. *Nanoscale Res. Lett.*, 1(1):84, 2006.

[83] S. Granick. Motions and relaxations of confined fluids. *Science*, 253(5026), 1374–1379, 1991.

[84] P. Gray and S. K. Scott. *Chemical Oscillations and Instabilities: Non-linear Chemical Kinetics*. Clarendon Press, Oxford, 1994.

[85] M. S. Green. Markoff random processes and the statistical mechanics of time-dependent phenomena. II. Irreversible processes in fluids. *Journal of Chemical Physics*, 22(3):398, 1954.

[86] D. J. Griffiths. *Introduction to Electrodynamics*. Pearson, San Francisco, 2008.

[87] D. J. Griffiths. *Introduction to Quantum Mechanics*. Pearson Education, Essex, 2014.

[88] L. J. Guo, X. Cheng, and C.-F. Chou. Fabrication of size-controllable nanoflu-idic channels by nanoimprinting and its application for DNA stretching. *Nano Lett.*, 4(1):69, 2004.

[89] A. Hanna, A. Saul, and K. Showalte. Detailed studies of propagating fronts in the iodate oxidation of arsenous acid. *J. Am. Chem. Soc.*, 104(14):3838, 1982.

[90] J. P. Hansen and I. R. McDonald. *Theory of Simple Liquids*. Academic Press, Amsterdam, 2006.

[91] J. S. Hansen. Where is the hydrodynamic limit? *Mol. Sim.*, 47(17), 1391–1401, 2021.

[92] J. S. Hansen, H. Bruus, B. D. Todd, and P. J Daivis. Rotational and spin vis-cosities of water: application to nanofluidics. *J. Chem. Phys.*, 133(14):144906, 2010.

[93] J. S. Hansen, P. J. Daivis, and B. D. Todd. Viscous properties of isotropic fluids composed of linear molecules: departure from the classical Navier-Stokes theory in nano-confined geometries. *Phys. Rev. E*, 80(4):046322, 2009.

[94] J. S. Hansen, P. J. Daivis, K. P. Travis, and B. D. Todd. Parameterization of the nonlocal viscosity kernel for an atomic fluid. *Phys. Rev. E*, 76(4):041121, 2007.

[95] J. S. Hansen, B. Nowakowski, and A. Lemarchand. Molecular dynamics simulations and master-equation description of a chemical wave front: effect of density and size of reaction zone on propagation speed. *J. Chem. Phys.*, 124(3):034503, 2006.

[96] J. S. Hansen and J. T. Ottesen. Molecular simulations of oscillatory flows in microfluidic channels. *Microfluid. Nanofluidics*, 2(4):301–307, 2006.

[97] J. S. Hansen. *Molecular Dynamics Simulations of Simple Reactive Mixtures*. PhD thesis, Roskilde University, 2003.

[98] J. S Hansen. Generalized extended Navier-Stokes theory: Multiscale spin relaxation in molecular fluids. *Phys. Rev. E*, 88(3):032101, 2013.

[99] J. S Hansen. Reduced dielectric response in spatially varying electric fields. *J. Chem. Phys.*, 143(19), 2015.

[100] J. S. Hansen. Multiscale dipole relaxation in dielectric materials. *Mol. Sim.*, 42(16):1364, 2016.

[101] J. S Hansen, P. J. Daivis, J. C. Dyre, B. D. Todd, and H. Bruus. Generalized extended Navier-Stokes theory: correlations in molecular fluids with intrinsic angular momentum. *J. Chem. Phys.*, 138(3):034503, 2013.

[102] J. S. Hansen, J. C. Dyre, P. J. Daivis, B. D. Todd, and H. Bruus. Continuum nanofluidics. *Langmuir*, 31(49):13275, 2015.

[103] J. S Hansen, M. L. Greenfield, and J. C. Dyre. Hydrodynamic relaxations in dissipative particle dynamics. *J. Chem. Phys.*, 148(3):034503, 2018.

[104] J. S Hansen, T. B Schrøder, and J. C. Dyre. Hydrodynamics of fluids approaching the viscous regime.

[105] J. S. Hansen, B. D. Todd, and P. J. Daivis. Prediction of fluid velocity slip at solid surfaces. *Phys. Rev. E*, 84(4):016313, 2011.

[106] K. Harris. Relations between the fractional Stokes-Einstein and Nernst-Einstein equations and velocity correlation coefficients in ionic liquids and molten salts. *J. Phys. Chem. B*, 114(29):9572, 2010.

[107] K.R. Harris, M. Kanakubo, N. Tsuchihashi, K. Ibuki, and M. Ueno. Effect of pressure on the transport properties of ionic liquids: 1-Alkyl-3-methylimidazolium salts. *J. Phys. Chem. B*, 112(32):9830, 2008.

[108] C. Herrero, T. Omori, Y. Yamaguchi, and L. Joly. Shear force measurement of the hydrodynamic wall position in molecular dynamics. *J. Chem. Phys.*, 151(4):041103.

[109] M. W. Hirsch, S. Smale, and R. L. Devaney. *Differential Equations, Dynamical Systems, and an Introduction to Chaos*. Elsevier, 2013.

[110] H. Hoang and G. Galliero. Shear viscosity of inhomogeneous fluids. *J. Chem. Phys.*, 136(12):124902, 2013.

[111] W. G. Hooever. *Smooth Particle Applied Mechanics*. World Scientific, Singapore, 2006.

[112] P. J. Hoogerbrugge and J. M. V. A. Koelman. Simulating microscopic hydrodynamics phenomena with dissipative particle dynamics. *Europhys. Lett.*, 19(3):155, 1992.

[113] J. Horbach and S. Succi. Lattice Boltzmann versus molecular dynamics simulation of nanoscale hydrodynamic flows. *Phys. Rev. Lett.*, 96(22):224503, 2006.

[114] R. G. Horn and J. N. Israelachvili. Direct measurement of structural forces between two surfaces in a nonpolar liquid. *J. Chem. Phys.*, 83(4):5311, 1981.

[115] K. Huang and I. Szlufarska. Green-Kubo relation for friction at liquid-solid interfaces. *Phys. Rev. E*, 89(3):032119, 2014.

[116] J. H. Irving and J. G. Kirkwood. The statistical mechanical theory of transport processes. IV. The equations of hydrodynamics. *J. Chem. Phys.*, 18(6):817–829, 1950.

[117] J. N. Israelachvili. *Intermolecular and Surface Forces*. Academic Press, London, 1992.

[118] C. Jackson and G. McKenna. The melting behavior of organic materials confined in porous solids. *J. Chem. Phys.*, 93(12):9002, 1990.

[119] L. Joly. Capillary filling with giant liquid/solid slip: dynamics of water uptake by carbon nanotubes. *J. Chem. Phys.*, 135(21):214705, 2011.

[120] L. Joly, C. Ybert, E. Trizac, and L. Bocquet. Hydrodynamics within the electric double layer on slipping surfaces. *Phys. Rev. Lett.*, 93(25):257805, 2004.

[121] L. Joly, C. Ybert, E. Trizac, and L. Bocquet. Liquid friction on charged surfaces: from hydrodynamic slippage to electrokinetics. *J. Chem. Phys.*, 125(20):204716, 2006.

[122] W. L. Jorgensen and J. Tirado-Rives The OPLS force field for proteins. Energy minimizations for crystals of cyclic peptides and crambin. *J. Am. Chem. Soc.*, 110(6):1657, 1988.

[123] S. K. Kannam, B. D. Todd, J. S. Hansen, and P. J. Daivis. Slip flow in graphene nanochannels. *J. Chem. Phys.*, 135(14):144701, 2011.

[124] S. K. Kannam, B. D. Todd, J. S. Hansen, and P. J. Daivis. Interfacial slip friction at a fluid-solid cylindrical boundary. *J. Chem. Phys.*, 136(24):244704, 2012.

[125] S. K. Kannam, B. D. Todd, J. S. Hansen, and P. J. Daivis. How fast does water flow in carbon nanochannels? *J. Chem. Phys.*, 138(9):094701, 2013.

[126] P. L. Kapitza. The study of heat transport in helium II. *J. Phys. (USSR)*, 4:181, 1941.

[127] G. Karniadakis, A. Beskok, and N. Aluru. *Microflows and Nanoflows: Fundamentals and Simulation*. Springer, New York, 2005.

[128] R. Karnik, K. Castelino, C. Duan, and A. Majumdar. Diffusion-limited patterning of molecules in nanofluidic channels. *Nano Lett. J. Phys. (USSR)*, 6(8):1736, 2006.

[129] T. R. Kirkpatrick, E. G. D. Cohen, and J. R. Dorfman. Light scattering by a fluid in a nonequilibrium steady state. I. Small gradients. *Phys. Rev. A*, 26(2):972, 1982.

[130] T. R. Kirkpatrick, E. G. D. Cohen, and J. R. Dorfman. Light scattering by a fluid in a nonequilibrium steady state. II. Large gradients. *Phys. Rev. A*, 26(2):995, 1982.

[131] S. Kjeldstrup and D. Bedaux. *Non-equilibrium Thermodynamics of Heterogenous Systems*. World Scientific Publishing, Singapore, 2008.

[132] C. Kleinstruer. *Microfluidics and Nanofluidics: Theory and Selected Applications*. John Wiley and Sons, New Jersey, 2015.

[133] S. Knudsen, J. C. Dyre, B. D. Todd, and J. S. Hansen. Hydrodynamic invariance. *J. Chem. Phys. Rev. E* 104(5), 054126, 2021.

[134] W. Kob and H. C. Andersen. Scaling Behavoir in the beta-relaxation regime of a supercooled Lennard-Jones mixture. *Phys. Rev. Lett.*, 73(10):1376, 1994.

[135] A. E. Kobryn and A. Kovalenko. Molecular theory of hydrodynamic boundary conditions in nanofluidics. *J. Chem. Phys.*, 129(13):134701–134717, 2008.

[136] J. Koplik, J. R. Banavar, and J. F. Willemsen. Molecular dynamics of fluid flow at solid surfaces. *Phys. Fluid. A*, 1(5):781–794, 1989.

[137] R. Kubo. Statistical-mechanical theory of irreversible processes. I. General theory and simple applications to magnetic and conduction problems. *Journal of the Physical Society of Japan*, 12(6):570–586, 1957.

[138] P. K. Kundu and I. M. Cohen. *Fluid Mechanics. 4th ed.*. Elsevier, Amsterdam, 2008.

[139] L. D. Landau and E. M. Lifshitz. *Statistical Physics*. 3rd edition. Pergamon Press, Oxford, 1980.

[140] L. D. Landau and E. M. Lifshitz. *Fluid Mechanics*. 2nd ed. Elsevier, Amsterdam, 1987.

[141] B. Lautrup. *Physics of Continuous Matter*. Institute of Physics Publishing, Bristol, 2005.

[142] R. A. Leach. *Molecular Modelling. Principles and Applications*. Prentice Hall, Harlow UK, 2001.

[143] L. Gary Leal. *Advanced Transport Phenomena*. Cambridge University Press, New York, 2007.

[144] S. D. Lecce and F. Bresme. Thermal polarization of water influences the thermoelectric response of aqueous solutions. *J. Phys. Chem. B*, 122(5):1662, 2018.

[145] A. W. Lees and S. F. Edwards. The computer study of transport processes under extreme conditions. *J. Phys. C: Solid State Phys.*, 5:1921, 1972.

[146] C. A. Lemarchand, P. J. Daivis, N. P. Bailey, B. D. Todd, and J. S. Hansen. Non-newtonian behavior and molecular structure of cooee bitumen under shear flow: a non-equilibrium molecular dynamics study. *J. Chem. Phys.*, 142(24):244501, 2015.

[147] D. Levesque, L. Verlet, and J. Kürkijarvi. Computer "experiment" on classical fluids. IV. Transport properties and time-correlation functions of the Lennard-Jones liquid near its triple point. *Phys. Rev. A*, 7(5):1690, 1973.

[148] D. Li, ed. *Encyclopedia of Microfluidics and Nanofluidics*. Springer-Verlag, Boston, 2008.

[149] S. De Luca, S. K. Kannam, B. D. Todd et al. Effects of confinement on the dielectric response of water extends up to mesoscale dimensions. *Langmuir*, 32(19):4765, 2016.

[150] S. De Luca, B. D. Todd, J. S. Hansen, and P. J. Daivis. Molecular dynamics study of nanoconfined water flow driven by rotating: electric fields under realistic experimental conditions. *Langmuir*, 30(11):3095, 2014.

[151] J. J. Magda, M. Tirrell, and H. T. Davis. Molecular dynamics of narrow, liquid-filled pores. *J. Chem. Phys.*, 83(4):1888, 1985.

[152] J. Mairhofer and R. J. Sadus. Thermodynamic properties of supercritical n-m Lennard-Jones fluids and isochoric and isobaric heat capacity maxima and minima. *J. Chem. Phys.*, 139(15):154503, 2013.

[153] M. Majumder, N. Chopra, R. Andrews, and B. J. Hinds. Nanoscale hydrodynamics: enhanced flow in carbon nanotubes. *Nature*, 438:44, 2005.

[154] C. A. Marsh, G. Backx, and M. H. Ernst. Static and dynamic properties of dissipative particle dynamics. *Phys. Rev. E*, 56(2):1676, 1997.

[155] M. G. Matin and J. I. Siepmann. Transferable atom description of n-alkanes. *J. Phys. Chem. B*, 102(14):2569, 1998.

[156] D. A. McQuarrie. *Statistical Mechanics*. Harper and Row, New York, 1976.

[157] A. A. Milischuk, V. Krewald, and B. M. Ladanyi. Water dynamics in silica nanopores: the self-intermediate scattering functions. *J. Phys. Chem.*, 136(22):224704, 2012.

[158] A. A. Milischuk and B. M. Ladanyi. Structure and dynamics of water confined in silica nanopores. *J. Phys. Chem.*, 135(17):174709, 2011.

[159] N. A. T. Miller, P. J. Daivis, I. K. Snook, and B. D. Todd. Computation of thermodynamic and transport properties to predict thermophoretic effects in an argon-krypton mixture. *J. Phys. Chem.*, 139(14):144504, 2013.

[160] R. J. D. Moore, J. S. Hansen, and B. D. Todd. Rotational viscosity of linear molecules: an equilibrium molecular dynamics study. *J. Phys. Chem.*, 128(22):224507, 2008.

[161] H. Mori. Transport, collective motion, and Brownian motion. *Progr. Theo. Phys.*, 33(3):423, 1965.

[162] J. D. Murray. *Mathematical Biology*. Springer Verlag, Berlin Heidelberg, 1989.

[163] C. L. M. H. Navier. Memoire sur les lois du mouvement des fluids. *Memoires de l'Academic Royale des Sciences de l'Institut de France*, 6:389–440, 1823.

[164] M. Nazari, A. Davoodabadi, D. Huang, T. Luo, and H. Ghasemi. Transport phenomena in nano/molecular confinements. *ACS Nano*, 14(12):16348, 2020.

[165] M. Neek-Amal, F. M. Peters, I. V. Grigoreva, and A. K. Geim. Commensurabilityeffects on viscosity of nanoconfined water. *ACS Nano*, 10(3):3685, 2016.

[166] R. R. Netz and H. Orland. Beyond Poisson-Boltzmann: fluctuation effects and correlation functions. *Euro. Phys. Journ. E*, 1:203, 2000.

[167] C. Nieto-Draghi, J. B. Avalos, and B. Rousseau. Computing the Soret coefficient in aqueous mixtures using boundary driven nonequilibrium molecular dynamics. *J. Chem. Phys.*, 122(11):114503, 2005.

[168] L. Onsager. Reciprocal relations in irreversible processes. I. *Phys. Rev.*, 37(4):405, 1931.

[169] U. R. Pedersen, T. Christensen, T. B. Schrøder, and J. C. Dyre. Feasibility of a single-parameter description of equilibrium viscous liquid dynamics. *Phys. Rev. E*, 77(1):011201, 2008.

[170] J. L. Perry and S. G. Kandlikar. Review of fabrication of nanochannels for single phase liquid flow. *Microfluid. Nanofluidics*, 2:185, 2006.

[171] J. Petravic and P. Harrowell. On the equilibrium calculation of the friction coefficient for liquid slip against a wall. *J. Chem. Phys.*, 127(17):174706–174711, 2007.

[172] J. Petravic and P. Harrowell. Erratum: On the equilibrium calculation of the friction coefficient for liquid slip against a wall [J. Chem. Phys. 127, 174706 (2007)]. *J. Chem. Phys.*, 128(20):209901, 2008.

[173] A. T. Pham, M. Barisik, and B. Kim. Pressure dependence of Kapitza resistance at gold/water and silicon/water interfaces. *J. Chem. Phys.*, 139(24):244702, 2013.

[174] N. Phan-Thien. *Understanding Viscoelasticity: an Introduction to Rheology*. Springer Verlag, Berlin, 2002.

[175] J. K. Platten. The Soret Effect: A review of recent experimental results. *J. Appl. Mech.*, 73(1):693, 2006.

[176] S. Prakash, A. Piruska, E. N. Gatimu et al. Nanofluidics: systems and applications. *IEEE Sensors J.*, 8(5):441, 2008.

[177] N. V. Priezjev. Rate-dependent slip boundary conditions for simple fluids. *Phys. Rev. E*, 75(5):051605, 2007.

[178] N. V. Priezjev. Molecular dynamics simulations of oscillatory couette flows with slip boundary conditions. *Microfluid. Nanofluidics*, 14:225, 2013.

[179] N. V. Priezjev and S. M. Troian. Influence of periodic wall roughness on the slip behaviour at liquid/solid interfaces: molecular-scale simulations versus continuum predictions. *J. Fluid Mech.*, 554:25, 2006.

[180] R. M. Puscasu, B. D. Todd, P. J. Daivis, and J. S Hansen. Nonlocal viscosity of polymer melts approaching their glassy state. *J. Chem. Phys.*, 133(14):144907, 2010.

[181] G. Raabe and R. J. Sadus. Influence of bond flexibility on the vapor-liquid phase equilibria of water. *J. Chem. Phys.*, 126(4):044701, 2007.

[182] A. Rahman. Correlations in the motion of atoms in liquid argon. *Phys. Rev.*, 136(2A):A405, 1967.

[183] M. A. Rahman and M. Z. Saghir. Thermodiffusion or soret effect: historical review. *International Journal of Heat and Mass Transfer*, 73:693–705, 2014.

[184] D. C. Rapaport. *The Art of Molecular Dynamics Simulation*. Cambridge University Press, Cambridge, 1995.

[185] R. L. Rowley and M. M. Painter. Diffusion and viscosity equation of state for a Lennard-Jones fluid obtained from molecular dynamics simulations. *Int. J. Therm. Phys.*, 18:1109, 1997.

[186] R. J. Sadus. *Molecular Simulation of Fluids. Theory, Algorithms and Object-Orientation*. Elsevier, Amsterdam, 1999.

[187] T. B. Schrøder and J. C. Dyre. Simplicity of condensed matter at its core: generic definition of a Roskilde-simple system. *J. Chem. Phys.*, 141(20):204502, 2014.

[188] E. Secchi, S. Marbach, A. Nigués et al., Massive radius-dependent flow slippage in carbon nanotubes. *Nature*, 537:210, 2016.

[189] D. J. Shaw. *Introduction to Colloid and Surface Chemistry*, 4th ed., 1992.

[190] R. F. Snider and K. S. Lewchuk. Irreversible thermodynamics of a fluid system with spin. *J. Chem. Phys.*, 46(8):3163, 1967.

[191] V. P. Sokhan and N. Quirke. Slip coefficient in nanoscale flow. *Phys. Rev. E*, 78(1):015301–015304, 2008.

[192] J. M. Thomas and W. J. McGaughey. *Heterogeneous Catalysis*. VCH Publishers Inc., New York, 1997.

[193] P. A. Thompson and M. O. Robbins. Shear flow near solids: epitaxial order and flow boundary conditions. *Phys. Rev. A*, 41(12):6830, 1990.

[194] P. A. Thompson and S. M. Troian. A general boundary condition for liquid flow at solid surfaces. *Nature*, 389:360–362, 1997.

[195] B. D. Todd and Peter J. Daivis. Homogeneous non-equilibrium molecular dynamics simulations of viscous flow: techniques and applications. *Mol. Simul.*, 33(3):189–229, 2007.

[196] B. D. Todd and D. Evans. Temperature profile for Poiseuille flow. *Phys. Rev. E*, 55(3):2800, 1997.

[197] B. D. Todd, D. J. Evans, and P. J. Daivis. Pressure tensor for inhomogenous fluids. *Phys. Rev. E*, 52(2):1627, 1995.

[198] B. D. Todd and J. S. Hansen. Nonlocal viscous transport and the effect on fluid stress. *Phys. Rev. E*, 78(5):051702, 2008.

[199] B. D. Todd, J. S. Hansen, and P. J. Daivis. Nonlocal shear stress for homogeneous fluids. *Phys. Rev. Lett.*, 100(19):195901–195904, 2008.

[200] B. D. Todd and P. J. Daivis. *Nonequilibrium Molecular Dynamics*. Cambridge University Press, Cambridge, 2017.

[201] S. Toxvaerd. The structure and thermodynamics of a solid-fluid interface. *J. Chem. Phys.*, 74(3):1998–2005, 1981.

[202] S. Toxvaerd and E. Præstgaard. Molecular dynamics calculation of the liquid structure up to a solid surface. *J. Chem. Phys.*, 67(11):5291, 1977.

[203] K. Trachenko and V. V. Brazhkin. Collective modes and thermodynamics of the liquid state. *Rep. Prog. Phys.*, 79(1):016502, 2016.

[204] G. P. Tolstov (translated by R. A. Silverman). *Fourier Series*. Dover Publications, Mineola, 1962.

[205] K. P. Travis, B. D. Todd, and D. J. Evans. Depature from Navier-Stokes hydrodynamics in confined liquids. *Phys. Rev. E*, 55(4):4288–4295, 1997.

[206] K. P. Travis, B. D. Todd, and D. J. Evans. Poiseuille flow of molecular fluids. *Physica A*, 240(1–2):315, 1997.

[207] D. J. Tritton. *Physical Fluid Dynamics*. Oxford Science Publications, Oxford, 1988.

[208] U. R. Pedersen, T. B. Schrøder, and J. C. Dyre Phase diagram of Kob-Andersen-Type binary Lennard-Jones mixtures.. *Phys. Rev. Lett*, 120(16):165501, 2018.

[209] S. Varghese, J. S. Hansen, and B. D. Todd. Improved methodology to compute the intrinsic friction coefficient at solid–liquid interfaces. *J. Chem. Phys.*, 154(18):184707, 2021.

[210] S. Varghese, S. K. Kannam, J. S. Hansen, and S. P. Sathian. Effect of hydrogen bonds on the dielectric properties of interfacial water. *Langmuir*, 35(24):8159, 2019.

[211] R. S. Voronov, D. V. Papavassiliou, and L. L. Lee. Boundary slip and wetting properties of interfaces: correlation of the contact angle with the slip length. *J. Chem. Phys.*, 124(20):204701, 2006.

[212] M. Whitby and N. Quirk. Fluid flow in carbon nanotubes and nanopipes. *Nature Nanotech.*, 2:87, 2007.

[213] Y. Wu, H. L. Tepper, and G. A. Voth. Flexible simple point-charge water model with improved liquid-state properties. *J. Chem. Phys.*, 124(2):024503, 2006.

[214] C. Yang, M. T. Dove, V. V. Brazhkin, and K. Trachenko. Emergence and evolution of the k gap in spectra of liquid and supercritical states. *Phys. Rev. Lett.*, 118(21–26):215502, 2017.

[215] C. Zhang. Note: On the dielectric constant of nanoconfined water. *J. Phys. Chem. Lett.*, 148(15):156101, 2018.

[216] J. Zhang, B. D. Todd, and K. P. Travis. Viscosity of confined inhomogeneous nonequilibrium fluids. *J. Chem. Phys.*, 121(21):10778–10786, 2004.

[217] J. Zhang, B. D. Todd, and K. P. Travis. Erratum: Viscosity of confined inhomogeneous nonequilibrium fluids. *J. Chem. Phys.*, 122(21):219901–219902, 2005.

[218] K. J. Zhang, M. E. Briggs, R. W. Gammon, and J. V. Sengers. Optical measurement of the Soret coefficient and the diffusion coefficient of liquid mixtures. *J. Chem. Phys.*, 104(17):6881, 1996.

[219] R. Zwanzig. Elementary derivation of time-correlation formulas for transport coefficients. *J. Chem. Phys.*, 40(9):2527, 1964.

Index